Teacher, Student, and Parent
One-Stop Internet Resources

Log on to
bookj.msscience.com

ONLINE STUDY TOOLS

- Section Self-Check Quizzes
- Interactive Tutor
- Chapter Review Tests
- Standardized Test Practice
- Vocabulary PuzzleMaker

ONLINE RESEARCH

- WebQuest Projects
- Prescreened Web Links
- Career Links
- Internet Labs

INTERACTIVE ONLINE STUDENT EDITION

- Complete Interactive Student Edition available at mhln.com

FOR TEACHERS

- Teacher Bulletin Board
- Teaching Today—Professional Development

SAFETY SYMBOLS

	HAZARD	EXAMPLES	PRECAUTION	REMEDY
DISPOSAL	Special disposal procedures need to be followed.	certain chemicals, living organisms	Do not dispose of these materials in the sink or trash can.	Dispose of wastes as directed by your teacher.
BIOLOGICAL	Organisms or other biological materials that might be harmful to humans	bacteria, fungi, blood, unpreserved tissues, plant materials	Avoid skin contact with these materials. Wear mask or gloves.	Notify your teacher if you suspect contact with material. Wash hands thoroughly.
EXTREME TEMPERATURE	Objects that can burn skin by being too cold or too hot	boiling liquids, hot plates, dry ice, liquid nitrogen	Use proper protection when handling.	Go to your teacher for first aid.
SHARP OBJECT	Use of tools or glassware that can easily puncture or slice skin	razor blades, pins, scalpels, pointed tools, dissecting probes, broken glass	Practice common-sense behavior and follow guidelines for use of the tool.	Go to your teacher for first aid.
FUME	Possible danger to respiratory tract from fumes	ammonia, acetone, nail polish remover, heated sulfur, moth balls	Make sure there is good ventilation. Never smell fumes directly. Wear a mask.	Leave foul area and notify your teacher immediately.
ELECTRICAL	Possible danger from electrical shock or burn	improper grounding, liquid spills, short circuits, exposed wires	Double-check setup with teacher. Check condition of wires and apparatus.	Do not attempt to fix electrical problems. Notify your teacher immediately.
IRRITANT	Substances that can irritate the skin or mucous membranes of the respiratory tract	pollen, moth balls, steel wool, fiberglass, potassium permanganate	Wear dust mask and gloves. Practice extra care when handling these materials.	Go to your teacher for first aid.
CHEMICAL	Chemicals can react with and destroy tissue and other materials	bleaches such as hydrogen peroxide; acids such as sulfuric acid, hydrochloric acid; bases such as ammonia, sodium hydroxide	Wear goggles, gloves, and an apron.	Immediately flush the affected area with water and notify your teacher.
TOXIC	Substance may be poisonous if touched, inhaled, or swallowed.	mercury, many metal compounds, iodine, poinsettia plant parts	Follow your teacher's instructions.	Always wash hands thoroughly after use. Go to your teacher for first aid.
FLAMMABLE	Flammable chemicals may be ignited by open flame, spark, or exposed heat.	alcohol, kerosene, potassium permanganate	Avoid open flames and heat when using flammable chemicals.	Notify your teacher immediately. Use fire safety equipment if applicable.
OPEN FLAME	Open flame in use, may cause fire.	hair, clothing, paper, synthetic materials	Tie back hair and loose clothing. Follow teacher's instruction on lighting and extinguishing flames.	Notify your teacher immediately. Use fire safety equipment if applicable.

 Eye Safety
Proper eye protection should be worn at all times by anyone performing or observing science activities.

 Clothing Protection
This symbol appears when substances could stain or burn clothing.

 Animal Safety
This symbol appears when safety of animals and students must be ensured.

 Handwashing
After the lab, wash hands with soap and water before removing goggles.

Astronomy

NATIONAL GEOGRAPHIC

bookj.msscience.com

Glencoe

New York, New York   Columbus, Ohio   Chicago, Illinois   Peoria, Illinois   Woodland Hills, California

Astronomy

This collection of images is of Jupiter, Io (one of its moons), Mars, and the Andromeda Galaxy. The Andromeda Galaxy is the most distant object visible to the human eye. At a distance of 2.2 million light years, it appears as a fuzzy patch of light in the night sky.

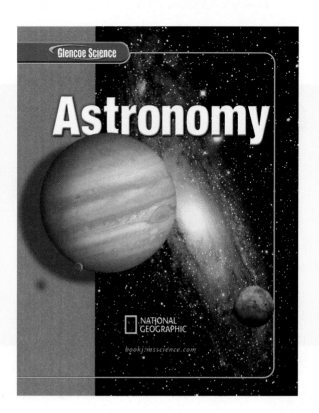

The McGraw-Hill Companies

Send all inquiries to:
Glencoe/McGraw-Hill
8787 Orion Place
Columbus, OH 43240-4027

ISBN: 0-07-861761-8

Printed in the United States of America.

5 6 7 8 9 10 027/043 09 08 07

Authors

Education Division
Washington, D.C.

Ralph M. Feather Jr., PhD
Science Department Chair
Derry Area School District
Derry, PA

Dinah Zike
Educational Consultant
Dinah-Might Activities, Inc.
San Antonio, TX

Series Consultants

CONTENT

William C. Keel, PhD
Department of Physics and Astronomy
University of Alabama
Tuscaloosa, AL

MATH

Teri Willard, EdD
Mathematics Curriculum Writer
Belgrade, MT

READING

Carol A. Senf, PhD
School of Literature, Communication, and Culture
Georgia Institute of Technology
Atlanta, GA

SAFETY

Aileen Duc, PhD
Science 8 Teacher
Hendrick Middle School, Plano ISD
Plano, TX

Sandra West, PhD
Department of Biology
Texas State University-San Marcos
San Marcos, TX

ACTIVITY TESTERS

Mary Helen Mariscal-Cholka
William D. Slider Middle School
El Paso, TX

Science Kit and Boreal Laboratories
Tonawanda, NY

Series Reviewers

Mary Ferneau
Westview Middle School
Goose Creek, SC

Annette D'Urso Garcia
Kearney Middle School
Commerce City, CO

Nerma Coats Henderson
Pickerington Lakeview Jr. High School
Pickerington, OH

Michael Mansour
Board Member
National Middle Level Science Teacher's Association
John Page Middle School
Madison Heights, MI

Mary Helen Mariscal-Cholka
William D. Slider Middle School
El Paso, TX

Sharon Mitchell
William D. Slider Middle School
El Paso, TX

HOW TO...

Use Your Science Book

Before You Read

- **Chapter Opener** Science is occurring all around you, and the opening photo of each chapter will preview the science you will be learning about. The **Chapter Preview** will give you an idea of what you will be learning about, and you can try the **Launch Lab** to help get your brain headed in the right direction. The **Foldables** exercise is a fun way to keep you organized.

- **Section Opener** Chapters are divided into two to four sections. The **As You Read** in the margin of the first page of each section will let you know what is most important in the section. It is divided into four parts. **What You'll Learn** will tell you the major topics you will be covering. **Why It's Important** will remind you why you are studying this in the first place! The **Review Vocabulary** word is a word you already know, either from your science studies or your prior knowledge. The **New Vocabulary** words are words that you need to learn to understand this section. These words will be in **boldfaced** print and highlighted in the section. Make a note to yourself to recognize these words as you are reading the section.

As You Read

- **Headings** Each section has a title in large red letters, and is further divided into blue titles and small red titles at the beginnings of some paragraphs. To help you study, make an outline of the headings and subheadings.

- **Margins** In the margins of your text, you will find many helpful resources. The **Science Online** exercises and **Integrate** activities help you explore the topics you are studying. **MiniLabs** reinforce the science concepts you have learned.

- **Building Skills** You also will find an **Applying Math** or **Applying Science** activity in each chapter. This gives you extra practice using your new knowledge, and helps prepare you for standardized tests.

- **Student Resources** At the end of the book you will find **Student Resources** to help you throughout your studies. These include **Science, Technology,** and **Math Skill Handbooks,** an **English/Spanish Glossary,** and an **Index.** Also, use your **Foldables** as a resource. It will help you organize information, and review before a test.

- **In Class** Remember, you can always ask your teacher to explain anything you don't understand.

FOLDABLES™
Study Organizer

Science Vocabulary Make the following Foldable to help you understand the vocabulary terms in this chapter.

STEP 1 **Fold** a vertical sheet of notebook paper from side to side.

STEP 2 **Cut** along every third line of only the top layer to form tabs.

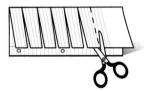

STEP 3 **Label** each tab with a vocabulary word from the chapter.

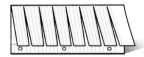

Build Vocabulary As you read the chapter, list the vocabulary words on the tabs. As you learn the definitions, write them under the tab for each vocabulary word.

Look For...

FOLDABLES™

At the beginning of every section.

In Lab

Working in the laboratory is one of the best ways to understand the concepts you are studying. Your book will be your guide through your laboratory experiences, and help you begin to think like a scientist. In it, you not only will find the steps necessary to follow the investigations, but you also will find helpful tips to make the most of your time.

- Each lab provides you with a **Real-World Question** to remind you that science is something you use every day, not just in class. This may lead to many more questions about how things happen in your world.

- Remember, experiments do not always produce the result you expect. Scientists have made many discoveries based on investigations with unexpected results. You can try the experiment again to make sure your results were accurate, or perhaps form a new hypothesis to test.

- Keeping a **Science Journal** is how scientists keep accurate records of observations and data. In your journal, you also can write any questions that may arise during your investigation. This is a great method of reminding yourself to find the answers later.

Look For...

- **Launch Labs** start every chapter.
- **MiniLabs** in the margin of each chapter.
- **Two Full-Period Labs** in every chapter.
- **EXTRA Try at Home Labs** at the end of your book.
- the **Web site** with laboratory demonstrations.

Before a Test

Admit it! You don't like to take tests! However, there *are* ways to review that make them less painful. Your book will help you be more successful taking tests if you use the resources provided to you.

- Review all of the **New Vocabulary** words and be sure you understand their definitions.

- Review the notes you've taken on your **Foldables,** in class, and in lab. Write down any question that you still need answered.

- Review the **Summaries** and **Self Check questions** at the end of each section.

- Study the concepts presented in the chapter by reading the **Study Guide** and answering the questions in the **Chapter Review.**

Look For...

- **Reading Checks** and **caption questions** throughout the text.
- the **Summaries** and **Self Check questions** at the end of each section.
- the **Study Guide** and **Review** at the end of each chapter.
- the **Standardized Test Practice** after each chapter.

Let's Get Started

To help you find the information you need quickly, use the Scavenger Hunt below to learn where things are located in Chapter 1.

1. What is the title of this chapter?

2. What will you learn in Section 1?

3. Sometimes you may ask, "Why am I learning this?" State a reason why the concepts from Section 2 are important.

4. What is the main topic presented in Section 2?

5. How many reading checks are in Section 1?

6. What is the Web address where you can find extra information?

7. What is the main heading above the sixth paragraph in Section 2?

8. There is an integration with another subject mentioned in one of the margins of the chapter. What subject is it?

9. List the new vocabulary words presented in Section 2.

10. List the safety symbols presented in the first Lab.

11. Where would you find a Self Check to be sure you understand the section?

12. Suppose you're doing the Self Check and you have a question about concept mapping. Where could you find help?

13. On what pages are the Chapter Study Guide and Chapter Review?

14. Look in the Table of Contents to find out on which page Section 2 of the chapter begins.

15. You complete the Chapter Review to study for your chapter test. Where could you find another quiz for more practice?

The Glencoe middle school science Student Advisory Board taking a timeout at COSI, a science museum in Columbus, Ohio.

Contents

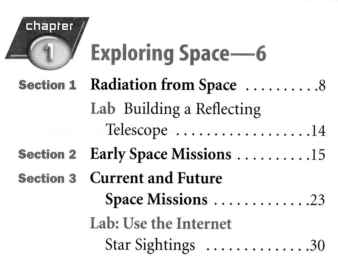

In each chapter, look for
these opportunities for
review and assessment:
• **Reading Checks**
• **Caption Questions**
• **Section Review**
• **Chapter Study Guide**
• **Chapter Review**
• **Standardized Test
Practice**
• **Online practice at
bookj.msscience.com**

Student Resources

Cross-Curricular Readings/Labs

available as a video lab

Design Your Own Labs

Model and Invent

Use the Internet Labs

Applying Math

Applying Science

INTEGRATE

Science Online

Standardized Test Practice

Life on Mars

Is there life on Mars? Ever since the 1600s, when scientists first looked at the sky with telescopes and determined that Mars is the most Earthlike planet in the solar system, they have asked this question.

In 1877, Italian scientist Giovanni Schiaparelli saw a network of lines on the surface of Mars and believed they were channels. Later, the American scientist Percival Lowell saw the same lines and claimed they were canals dug by martians.

Today scientists know that flowing water created many of the martian surface features. But scientists still wonder whether simple life-forms existed on Mars or might even exist today. To answer this question, they began undertaking space missions. One objective of the missions is to gather information on whether Mars has or ever had the conditions necessary for life, such as the presence of flowing water.

In 1964, scientists sent a space probe to take photographs of Mars. Examining the photos, they decided the planet is too cold and dry for life. Later probes showed that Mars might have been warm and wet billions of years ago. However, scientists still thought it had been cold and dry since those times.

Figure 1 This martian rock fell to Earth (Antarctica) as a meteorite.

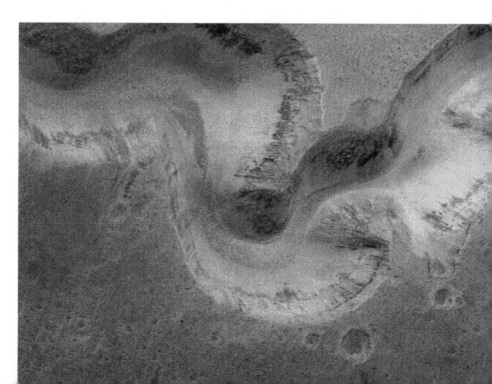

Figure 2 Scientists now know that the lines on Mars's surface were created by flowing water.

Then, in June 2000, scientists made an astounding discovery. Photographs taken by a new space probe showed evidence of recent erosion by running water. But if Mars is so cold, how could liquid water exist?

Further study of the new photographs convinced scientists that lava had flowed on Mars in the recent past. This means that Mars's interior must be warmer than previously thought. This heat could melt underground ice and allow it to flow to the surface as liquid water. Liquid water could help support life.

In 1984, scientists in Antarctica found a martian meteorite—a small piece of rock that was blasted into space when Mars was hit by a much larger meteorite. When scientists examined the meteorite with microscopes, they discovered strange shapes inside.

Similar shapes have been found in Earth's rocks and are thought to be the fossilized remains of bacteria that lived billions of years ago. Some scientists thought that the shapes in the martian meteorite were fossils of tiny forms of martian life.

Others thought the shapes were only globules of minerals formed when water changed the rocks on Mars. To test this idea, scientists tried to reproduce the shapes in a laboratory. When their experiment was completed, they saw globules of minerals like those in the martian meteorite. They concluded that the shapes probably were not fossils of martian life-forms.

Figure 3 The darker areas in this photograph are newer lava flows that broke up along their edges.

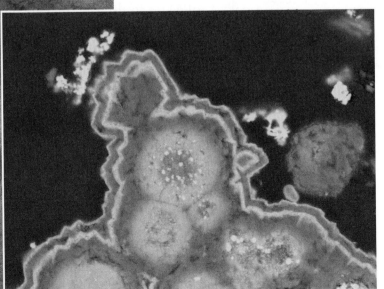

Figure 4 These globules of minerals made in a lab look like the shapes inside the martian meteorite.

Science

Trying to find out whether life ever existed on Mars is just one example of doing science. Science is the process of observing, experimenting, and thinking about the universe to create knowledge. In fact, the word *science* comes from the Latin word *scientia*, which means knowledge. Every time you answer a question by observing the world or testing an idea, you are doing science.

The Earth sciences study Earth—its land, oceans, and atmosphere—as well as other objects in the universe. In this book, you will learn about astronomy, the study of outer space.

Figure 5 Scientists use scientific methods to answer questions about life on Mars.

Scientific Methods

1. **Identify a question.**
   Determine a question to be answered.
2. **Form a hypothesis.**
   Gather information and propose an answer to the question.
3. **Test the hypothesis.**
   Perform experiments or make observations to see if the hypothesis is supported.
4. **Analyze results.**
   Look for patterns in the data that have been collected.
5. **Draw a conclusion.**
   Decide what the test results mean. Communicate your results.

Scientific Methods

Many scientists are working to answer the question of whether life ever existed on Mars. These scientists use a variety of methods to try to answer the question. These methods are commonly called scientific methods. **Scientific methods** are procedures used to investigate a question scientifically.

Identifying a Question

The first step in doing science is identifying a question. One such question is *Did life ever exist on Mars?* Answering this question could lead to many others. Scientists might want to know under what kinds of conditions life can survive. They also might want to ask whether the surface of Mars could have met such conditions. If you have ever participated in a science fair, you had to identify a question before you began your project.

Forming a Hypothesis

The next step is to gather information about the question and form a hypothesis. You can find information by going to the library and reading books or magazines, by using the Internet, or by talking to other people about the question. A **hypothesis** is a possible answer to a question. One hypothesis about the shapes in the martian meteorite is that *the shapes are fossils of tiny life-forms that lived on Mars long ago.* Another hypothesis is that *the shapes are globules of minerals formed inside martian rocks.*

Testing the Hypothesis

To find out whether a hypothesis is correct, scientists must test it. They do this by performing experiments or making observations. When scientists tried to produce globules of minerals that looked like the shapes in the martian meteorite, they were testing their hypothesis.

Analyzing Results

As scientists perform tests, they collect lots of information, or **data,** that must be analyzed. Data about the martian meteorite include measurements, microscope photographs, and chemical studies of the strange shapes. The test data must be organized and studied. Many times scientists make graphs so they can see patterns in the data. They also use computers to check the data.

Figure 6 Scientists often use microscopes and other equipment to test hypotheses.

Drawing a Conclusion

Often, the last step in a scientific method is to draw a conclusion. In this step, scientists decide what the results of their tests and observations mean. Sometimes the original hypothesis is not supported by the data. When this happens, the scientists begin again with a new hypothesis. Other times, though, the original hypothesis is supported. If a hypothesis is supported by repeatable experiments and many observations over time, it could become a theory. In science, a **theory** is an idea that has been tested and can explain a large set of observations. For instance, the claim that liquid water has, at some time, flowed over the martian surface is a theory. It might be many years before scientists know whether any of the hypotheses about the martian meteorite and life on Mars are correct.

In recent years, scientists have discovered microscopic organisms living kilometers beneath the surface of Earth. Some scientists have hypothesized that simple life-forms might exist deep below the surface of Mars, too. Describe one way that scientists could test this hypothesis.

Exploring Space

Fiery end or new beginning?

These colorful streamers are the remains of a star that exploded in a nearby galaxy thousands of years ago. Eventually, new stars and planets may form from this material, just as our Sun and planets formed from similar debris billions of years ago.

Science Journal Do you think space exploration is worth the risk and expense? Explain why.

Start-Up Activities

An Astronomer's View

You might think exploring space with a telescope is easy because the stars seem so bright and space is dark. But starlight passing through Earth's atmosphere, and differences in temperature and density of the atmosphere can distort images.

1. Cut off a piece of clear plastic wrap about 15 cm long.

2. Place an opened book in front of you and observe the clarity of the text.

3. Hold the piece of plastic wrap close to your eyes, keeping it taut using both hands.

4. Look at the same text through the plastic wrap.

5. Fold the plastic wrap in half and look at the text again through both layers.

6. **Think Critically** Write a paragraph in your Science Journal comparing reading text through plastic wrap to an astronomer viewing stars through Earth's atmosphere. Predict what might occur if you increased the number of layers.

Preview this chapter's content and activities at bookj.msscience.com

Exploring Space Make the following Foldable to help identify what you already know, what you want to know, and what you learned about exploring space.

STEP 1 **Fold** a vertical sheet of paper from side to side with the front edge about 1.25 cm shorter than the back.

STEP 2 **Turn** lengthwise and **fold** into thirds.

STEP 3 **Unfold and cut** only the top layer along both folds to make three tabs. **Label** each tab.

Know? | Like to know? | Learned?

Identify Questions Before you read the chapter, write what you already know about exploring space under the left tab of your Foldable, and write questions about what you'd like to know under the center tab. After you read the chapter, list what you learned under the right tab.

Radiation from Space

as you read

What You'll Learn

- **Explain** the electromagnetic spectrum.
- **Identify** the differences between refracting and reflecting telescopes.
- **Recognize** the differences between optical and radio telescopes.

Why It's Important

Learning about space can help us better understand our own world.

Review Vocabulary

telescope: an instrument that can magnify the size of distant objects

New Vocabulary

- electromagnetic spectrum
- refracting telescope
- reflecting telescope
- observatory
- radio telescope

Electromagnetic Waves

As you have read, we have begun to explore our solar system and beyond. With the help of telescopes like the *Hubble,* we can see far into space, but if you've ever thought of racing toward distant parts of the universe, think again. Even at the speed of light it would take many years to reach even the nearest stars.

Light from the Past When you look at a star, the light that you see left the star many years ago. Although light travels fast, distances between objects in space are so great that it sometimes takes millions of years for the light to reach Earth.

The light and other energy leaving a star are forms of radiation. Radiation is energy that is transmitted from one place to another by electromagnetic waves. Because of the electric and magnetic properties of this radiation, it's called electromagnetic radiation. Electromagnetic waves carry energy through empty space and through matter.

Electromagnetic radiation is everywhere around you. When you turn on the radio, peer down a microscope, or have an X ray taken—you're using various forms of electromagnetic radiation.

Figure 1 The electromagnetic spectrum ranges from gamma rays with wavelengths of less than 0.000 000 000 01 m to radio waves more than 100,000 m long. **Observe** how frequency changes as wavelength shortens.

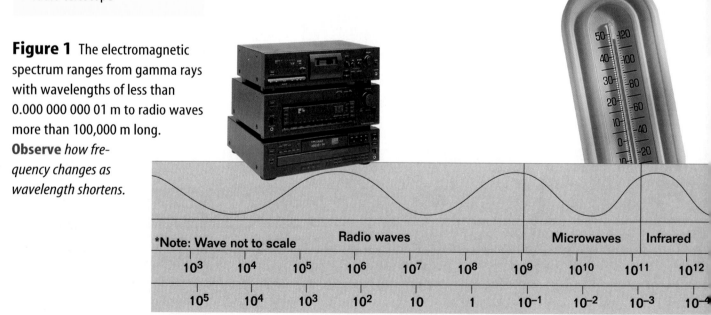

*Note: Wave not to scale	Radio waves						Microwaves		Infrared
10^3	10^4	10^5	10^6	10^7	10^8	10^9	10^{10}	10^{11}	10^{12}
10^5	10^4	10^3	10^2	10	1	10^{-1}	10^{-2}	10^{-3}	10^{-4}

Electromagnetic Radiation

Sound waves, which are a type of mechanical wave, can't travel through empty space. How, then, do we hear the voices of the astronauts while they're in space? When astronauts speak into a microphone, the sound waves are converted into electromagnetic waves called radio waves. The radio waves travel through space and through Earth's atmosphere. They're then converted back into sound waves by electronic equipment and audio speakers.

Radio waves and visible light from the Sun are just two types of electromagnetic radiation. Other types include gamma rays, X rays, ultraviolet waves, infrared waves, and microwaves. **Figure 1** shows these forms of electromagnetic radiation arranged according to their wavelengths. This arrangement of electromagnetic radiation is called the **electromagnetic spectrum.** Forms of electromagnetic radiation also differ in their frequencies. Frequency is the number of wave crests that pass a given point per unit of time. The shorter the wavelength is, the higher the frequency, as shown in **Figure 1.**

Speed of Light

Although the various electromagnetic waves differ in their wavelengths, they all travel at 300,000 km/s in a vacuum. This is called the speed of light. Visible light and other forms of electromagnetic radiation travel at this incredible speed, but the universe is so large that it takes millions of years for the light from some stars to reach Earth.

When electromagnetic radiation from stars and other objects reaches Earth, scientists use it to learn about its source. One tool for studying such electromagnetic radiation is a telescope.

INTEGRATE Health

Ultraviolet Light Many newspapers include an ultraviolet (UV) index to urge people to minimize their exposure to the Sun. Compare the wavelengths and frequencies of red and violet light, shown below in **Figure 1.** Infer what properties of UV light cause damage to tissues of organisms.

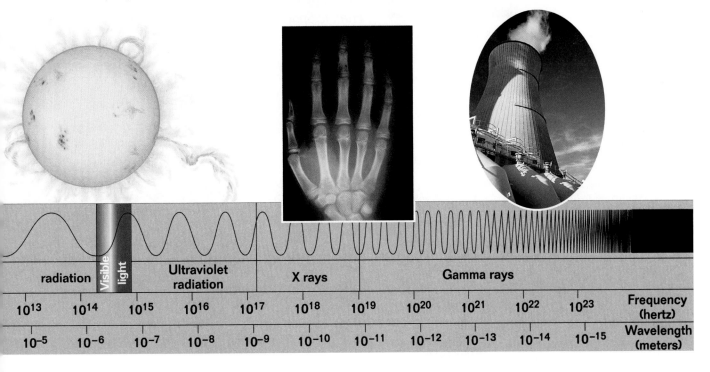

radiation	Visible light	Ultraviolet radiation	X rays		Gamma rays			

10^{13}	10^{14}	10^{15}	10^{16}	10^{17}	10^{18}	10^{19}	10^{20}	10^{21}	10^{22}	10^{23}	**Frequency (hertz)**
10^{-5}	10^{-6}	10^{-7}	10^{-8}	10^{-9}	10^{-10}	10^{-11}	10^{-12}	10^{-13}	10^{-14}	10^{-15}	**Wavelength (meters)**

Optical Telescopes

Optical telescopes use light, which is a form of electromagnetic radiation, to produce magnified images of objects. Light is collected by an objective lens or mirror, which then forms an image at the focal point of the telescope. The focal point is where light that is bent by the lens or reflected by the mirror comes together to form an image. The eyepiece lens then magnifies the image. The two types of optical telescopes are shown in **Figure 2.**

A **refracting telescope** uses convex lenses, which are curved outward like the surface of a ball. Light from an object passes through a convex objective lens and is bent to form an image at the focal point. The eyepiece magnifies the image.

A **reflecting telescope** uses a curved mirror to direct light. Light from the object being viewed passes through the open end of a reflecting telescope. This light strikes a concave mirror, which is curved inward like a bowl and located at the base of the telescope. The light is reflected off the interior surface of the bowl to the focal point where it forms an image. Sometimes, a smaller mirror is used to reflect light into the eyepiece lens, where it is magnified for viewing.

Figure 2 These diagrams show how each type of optical telescope collects light and forms an image.

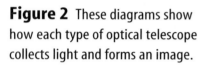

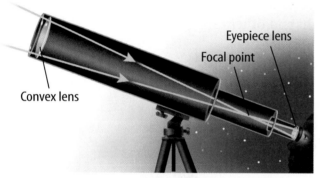

In a refracting telescope, a convex lens focuses light to form an image at the focal point.

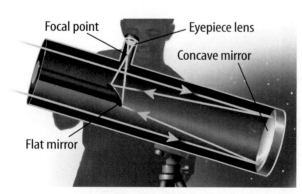

In a reflecting telescope, a concave mirror focuses light to form an image at the focal point.

Using Optical Telescopes

Most optical telescopes used by professional astronomers are housed in buildings called **observatories.** Observatories often have dome-shaped roofs that can be opened up for viewing. However, not all telescopes are located in observatories. The *Hubble Space Telescope* is an example.

Optical telescopes are widely available for use by individuals.

Hubble Space Telescope The *Hubble Space Telescope* was launched in 1990 by the space shuttle *Discovery*. Because *Hubble* is located outside Earth's atmosphere, which absorbs and distorts some of the energy received from space, it should have produced clear images. However, when the largest mirror of this reflecting telescope was shaped, a mistake was made. As a result, images obtained by the telescope were not as clear as expected. In December 1993, a team of astronauts repaired the *Hubble Space Telescope* by installing a set of small mirrors designed to correct images obtained by the faulty mirror. Two more missions to service *Hubble* were carried out in 1997 and 1999, shown in **Figure 3.** Among the objects viewed by *Hubble* after it was repaired in 1999 was a large cluster of galaxies known as Abell 2218.

Reading Check *Why is* **Hubble** *located outside Earth's atmosphere?*

Figure 3 The *Hubble Space Telescope* was serviced at the end of 1999. Astronauts replaced devices on *Hubble* that are used to stabilize the telescope.

Procedure

1. Obtain a **cardboard tube** from an empty roll of paper towels.
2. Go outside on a clear night about two hours after sunset. Look through the cardboard tube at a specific constellation decided upon ahead of time.
3. Count the number of stars you can see without moving the observing tube. Repeat this three times.
4. Calculate the average number of observable stars at your location.

Analysis

1. Compare and contrast the number of stars visible from other students' homes.
2. Explain the causes and effects of your observations.

Large Reflecting Telescopes Since the early 1600s, when the Italian scientist Galileo Galilei first turned a telescope toward the stars, people have been searching for better ways to study what lies beyond Earth's atmosphere. For example, the twin Keck reflecting telescopes, shown in **Figure 4,** have segmented mirrors 10 m wide. Until 2000, these mirrors were the largest reflectors ever used. To cope with the difficulty of building such huge mirrors, the Keck telescope mirrors are built out of many small mirrors that are pieced together. In 2000, the European Southern Observatory's telescope, in Chile, consisted of four 8.2-m reflectors, making it the largest optical telescope in use.

Reading Check *About how long have people been using telescopes?*

Active and Adaptive Optics The most recent innovations in optical telescopes involve active and adaptive optics. With active optics, a computer corrects for changes in temperature, mirror distortions, and bad viewing conditions. Adaptive optics is even more ambitious. Adaptive optics uses a laser to probe the atmosphere and relay information to a computer about air turbulence. The computer then adjusts the telescope's mirror thousands of times per second, which lessens the effects of atmospheric turbulence. Telescope images are clearer when corrections for air turbulence, temperature changes, and mirror-shape changes are made.

Figure 4 The twin Keck telescopes on Mauna Kea in Hawaii can be used together, more than doubling their ability to distinguish objects. A Keck reflector is shown in the inset photo. Currently, plans include using these telescopes, along with four others to obtain images that will help answer questions about the origin of planetary systems.

Radio Telescopes

As shown in the spectrum illustrated in **Figure 1,** stars and other objects radiate electromagnetic energy of various types. Radio waves are an example of long-wavelength energy in the electromagnetic spectrum. A **radio telescope,** such as the one shown in **Figure 5,** is used to study radio waves traveling through space. Unlike visible light, radio waves pass freely through Earth's atmosphere. Because of this, radio telescopes are useful 24 hours per day under most weather conditions.

Radio waves reaching Earth's surface strike the large, concave dish of a radio telescope. This dish reflects the waves to a focal point where a receiver is located. The information allows scientists to detect objects in space, to map the universe, and to search for signs of intelligent life on other planets.

Figure 5 This radio telescope is used to study radio waves traveling through space.

section 1 review

Summary

Electromagnetic Waves
- Light is a form of electromagnetic radiation.
- Electromagnetic radiation includes radio waves, microwaves, X rays, gamma rays, and infrared and ultraviolet radiation.
- Light travels at 300,000 km/s in a vacuum.

Optical Telescopes
- A refracting telescope uses lenses to collect, focus, and view light.
- A reflecting telescope uses a mirror to collect and focus light and a lens to view the image.
- Computers and lasers are used to reduce problems caused by looking through Earth's atmosphere.
- These telescopes are housed in domed buildings called observatories.
- Placing a telescope in space avoids problems caused by Earth's atmosphere.

Radio Telescopes
- Radio telescopes collect and measure radio waves coming from stars and other objects.

Self Check

1. **Identify** one advantage of radio telescopes over optical telescopes.
2. **Infer** If red light has a longer wavelength than blue light, which has a greater frequency?
3. **Explain** the difference between sound waves and radio waves.
4. **Describe** how adaptive optics in a telescope help solve problems caused by atmospheric turbulence.
5. **Think Critically** It takes light from the closest star to Earth (other than the Sun) about four years to reach Earth. If intelligent life were on a planet circling that star, how long would it take for scientists on Earth to send them a radio transmission and for the scientists to receive their reply?

Applying Math

6. **Calculate** how long it takes for a radio signal to reach the Moon, which is about 380,000 km away.
7. **Use Numbers** If an X ray has a frequency of 10^{18} hertz and a gamma ray has a frequency of 10^{21} hertz, how many times greater is the frequency of the gamma ray?

Building a Reflecting Telescope

Nearly four hundred years ago, Galileo Galilei saw what no human had ever seen. Using the telescope he built, he saw moons around Jupiter, details of lunar craters, and sunspots. What was it like to make these discoveries? Find out as you make your own reflecting telescope.

◉ Real-World Question

How do you construct a reflecting telescope?

Goals

- **Construct** a reflecting telescope.
- **Observe** magnified images using the telescope and different magnifying lenses.

Materials

flat mirror
shaving or cosmetic mirror (a curved, concave mirror)
magnifying lenses of different magnifications (3–4)

Safety Precautions

WARNING: *Never observe the Sun directly or with mirrors.*

◉ Procedure

1. Position the cosmetic mirror so that you can see the reflection of the object you want to look at. Choose an object such as the Moon, a planet, or an artificial light source.
2. Place the flat mirror so that it is facing the cosmetic mirror.
3. Adjust the position of the flat mirror until

you can see the reflection of the object in it.
4. View the image of the object in the flat mirror with one of your magnifying lenses. Observe how the lens magnifies the image.
5. Use your other magnifying lenses to view the image of the object in the flat mirror. Observe how the different lenses change the image of the object.

◉ Analyze Your Data

1. **Describe** how the image changed when you used different magnifying lenses.
2. **Identify** the part or parts of your telescope that reflected the light of the image.
3. **Identify** the parts of your telescope that magnified the image.

◉ Conclude and Apply

1. **Explain** how the three parts of your telescope worked to reflect and magnify the light of the object.
2. **Infer** how the materials you used would have differed if you had constructed a refracting instead of a reflecting telescope.

Communicating Your Data

Write an instructional pamphlet for amateur astronomers about how to construct a reflecting telescope.

Early Space Missions

The First Missions into Space

You're offered a choice—front-row-center seats for this weekend's rock concert, or a copy of the video when it's released. Wouldn't you rather be right next to the action? Astronomers feel the same way about space. Even though telescopes have taught them a great deal about the Moon and planets, they want to learn more by going to those places or by sending spacecraft where humans can't go.

Rockets The space program would not have gotten far off the ground using ordinary airplane engines. To break free of gravity and enter Earth's orbit, spacecraft must travel at speeds greater than 11 km/s. The space shuttle and several other spacecrafts are equipped with special engines that carry their own fuel. **Rockets,** like the one in **Figure 6,** are engines that have everything they need for the burning of fuel. They don't even require air to carry out the process. Therefore, they can work in space, which has no air. The simplest rocket engine is made of a burning chamber and a nozzle. More complex rockets have more than one burning chamber.

Rocket Types Two types of rockets are distinguished by the type of fuel they use. One type is the liquid-propellant rocket and the other is the solid-propellant rocket. Solid-propellant rockets are generally simpler but they can't be shut down after they are ignited. Liquid-propellant rockets can be shut down after they are ignited and can be restarted. For this reason, liquid-propellant rockets are preferred for use in long-term space missions. Scientists on Earth can send signals that start and stop the spacecraft's engines whenever they want to modify its course or adjust its orbit. Liquid propellants successfully powered many space probes, including the two *Voyagers* and *Galileo.*

as you read

What You'll Learn

- **Compare and contrast** natural and artificial satellites.
- **Identify** the differences between artificial satellites and space probes.
- **Explain** the history of the race to the Moon.

Why It's Important

Early missions that sent objects and people into space began a new era of human exploration.

Review Vocabulary
thrust: the force that propels an aircraft or missile

New Vocabulary
- rocket
- satellite
- orbit
- space probe
- Project Mercury
- Project Gemini
- Project Apollo

Figure 6 Rockets differ according to the types of fuel used to launch them. Liquid oxygen is used often to support combustion.

Figure 7 The space shuttle uses both liquid and solid fuels. Here the red liquid fuel tank is visible behind a white, solid rocket booster.

Rocket Launching Solid-propellant rockets use a rubberlike fuel that contains its own oxidizer. The burning chamber of a rocket is a tube that has a nozzle at one end. As the solid propellant burns, hot gases exert pressure on all inner surfaces of the tube. The tube pushes back on the gas except at the nozzle where hot gases escape. Thrust builds up and pushes the rocket forward.

Liquid-propellant rockets use a liquid fuel and an oxidizer, such as liquid oxygen, stored in separate tanks. To ignite the rocket, the oxidizer is mixed with the liquid fuel in the burning chamber. As the mixture burns, forces are exerted and the rocket is propelled forward. **Figure 7** shows the space shuttle, with both types of rockets, being launched.

Applying Math — Make and Use Graphs

DRAWING BY NUMBERS Points are defined by two coordinates, called an ordered pair. To plot an ordered pair, find the first number on the horizontal *x*-axis and the second on the vertical *y*-axis. The point is placed where these two coordinates intersect. Line segments are drawn to connect points. Draw a symmetrical house by using an *x-y* grid and these coordinates: (1,1), (5,1), (5,4), (3,6), (1,4)

Solution

1 *On a piece of graph paper, label and number the x-axis 0 to 6 and the y-axis 0 to 6, as shown here.*

2 *Plot the above points and connect them with straight line segments, as shown here.*

Section	Points
1	(1, −8) (3, −13) (6, −21) (9, −21) (9, −17) (8, −15) (8, −12) (6, −8) (5, −4) (4, −3) (4, −1) (5,1) (6,3) (8,3) (9,4) (9,7) (7,11) (4,14) (4,22) (−9,22) (−9,10) (−10,5) (−11, −1) (−11, −7) (−9, −8) (−8, −7) (−8, −1) (−6,3) (−6, −3) (−6, −9) (−7, −20) (−8, −21) (−4, −21) (−4, −18) (−3, −14) (−1, −8)
2	(0,11) (2,13) (2,17) (0,19) (−4,19) (−6,17) (−6,13) (−4,11)
3	(−4,9) (1,9) (1,5) (−1,5) (−2,6) (−4,6)

Practice Problems

1. Label and number the *x*-axis −12 to 10 and the *y*-axis −22 to 23. Draw an astronaut by plotting and connecting the points in each section. Do not draw segments to connect points in different sections.

2. Make your own drawing on graph paper and write its coordinates as ordered pairs. Then give it to a classmate to solve.

 For more practice, visit bookj.msscience.com/math_practice

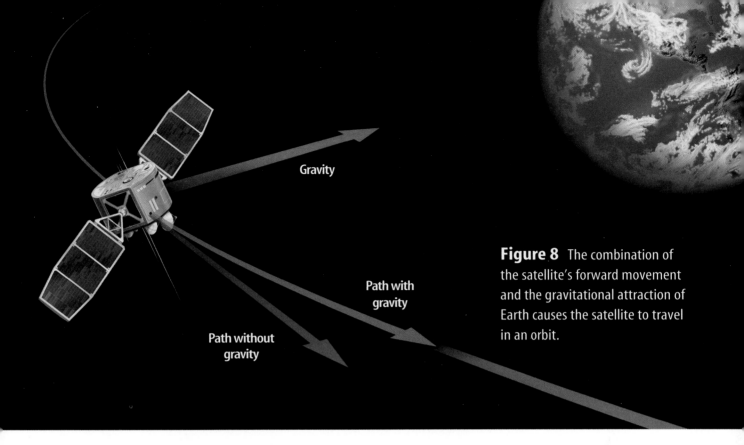

Gravity

Path with gravity

Path without gravity

Figure 8 The combination of the satellite's forward movement and the gravitational attraction of Earth causes the satellite to travel in an orbit.

Satellites The space age began in 1957 when the former Soviet Union used a rocket to send *Sputnik I* into space. *Sputnik I* was the first artificial satellite. A **satellite** is any object that revolves around another object. When an object enters space, it travels in a straight line unless a force, such as gravity, makes it turn. Earth's gravity pulls a satellite toward Earth. The result of the satellite traveling forward while at the same time being pulled toward Earth is a curved path, called an **orbit,** around Earth. This is shown in **Figure 8.** *Sputnik I* orbited Earth for 57 days before gravity pulled it back into the atmosphere, where it burned up.

Figure 9 Data obtained from the satellite *Terra,* launched in 1999, illustrates the use of space technology to study Earth. This false-color image includes data on spring growth, sea-surface temperature, carbon monoxide concentrations, and reflected sunlight, among others.

Satellite Uses *Sputnik I* was an experiment to show that artificial satellites could be made and placed into orbit around Earth.

Today, thousands of artificial satellites orbit Earth. Communication satellites transmit radio and television programs to locations around the world. Other satellites gather scientific data, like those shown in **Figure 9,** which can't be obtained from Earth, and weather satellites constantly monitor Earth's global weather patterns.

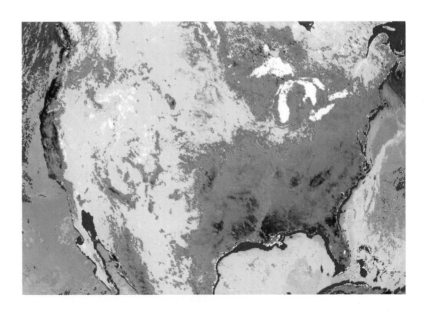

INTEGRATE
Career

Astronomy Astronomers today have more choices than ever before. Some still use optical telescopes to study stars and galaxies. Others explore the universe using the radio, X-ray, infrared, or even gamma-ray regions of the electromagnetic spectrum. Still others deal with theory and work with physicists to understand the big bang and the nature of matter in the universe. Government, universities, and private industry offer jobs for astronomers.

Space Probes

Not all objects carried into space by rockets become satellites. Rockets also can be used to send instruments into space to collect data. A **space probe** is an instrument that gathers information and sends it back to Earth. Unlike satellites that orbit Earth, space probes travel into the solar system as illustrated in **Figure 10.** Some even have traveled to the edge of the solar system. Among these is *Pioneer 10,* launched in 1972. Although its transmitter failed in 2003, it continues on through space. Also, both *Voyager* spacecrafts should continue to return data on the outer reaches of the solar system until about 2020.

Space probes, like many satellites, carry cameras and other data-gathering equipment, as well as radio transmitters and receivers that allow them to communicate with scientists on Earth. **Table 1** shows some of the early space probes launched by the National Aeronautics and Space Administration (NASA).

Table 1 Some Early Space Missions				
Mission Name		**Date Launched**	**Destination**	**Data Obtained**
Mariner 2		August 1962	Venus	verified high temperatures in Venus's atmosphere
Pioneer 10		March 1972	Jupiter	sent back photos of Jupiter—first probe to encounter an outer planet
Viking 1		August 1975	Mars	orbiter mapped the surface of Mars; lander searched for life on Mars
Magellan		May 1989	Venus	mapped Venus's surface and returned data on the composition of Venus's atmosphere

Figure 10

Probes have taught us much about the solar system. As they travel through space, these car-size craft gather data with their onboard instruments and send results back to Earth via radio waves. Some data collected during these missions are made into pictures, a selection of which is shown here.

Mariner 10

Mercury

A In 1974, *Mariner 10* obtained the first good images of the surface of Mercury.

Venera 8

B A Soviet *Venera* probe took this picture of the surface of Venus on March 1, 1982. Parts of the spacecraft's landing gear are visible at the bottom of the photograph.

Magellan

D In 1990, *Magellan* imaged craters, lava domes, and great rifts, or cracks, on the surface of Venus.

Venus

Neptune

Voyager 2

C The *Voyager 2* mission included flybys of the outer planets Jupiter, Saturn, Uranus, and Neptune. *Voyager* took this photograph of Neptune in 1989 as the craft sped toward the edge of the solar system.

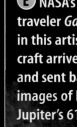

Jupiter

Galileo

E NASA's veteran space traveler *Galileo* nears Jupiter in this artist's drawing. The craft arrived at Jupiter in 1995 and sent back data, including images of Europa, one of Jupiter's 61 moons, seen below in a color-enhanced view.

Europa

***Voyager* and *Pioneer* Probes** Space probes *Voyager 1* and *Voyager 2* were launched in 1977 and now are heading toward deep space. *Voyager 1* flew past Jupiter and Saturn. *Voyager 2* flew past Jupiter, Saturn, Uranus, and Neptune. These probes will explore beyond the solar system as part of the Voyager Interstellar Mission. Scientists expect these probes to continue to transmit data to Earth for at least 20 more years.

Pioneer 10, launched in 1972, was the first probe to survive a trip through the asteroid belt and encounter an outer planet, Jupiter. As of 2003, *Pioneer 10* was more than 12 billion km from Earth, and will continue beyond the solar system. The probe carries a gold medallion with an engraving of a man, a woman, and Earth's position in the galaxy.

Galileo Launched in 1989, *Galileo* reached Jupiter in 1995. In July 1995, *Galileo* released a smaller probe that began a five-month approach to Jupiter. The small probe took a parachute ride through Jupiter's violent atmosphere in December 1995.

Before being crushed by the atmospheric pressure, it transmitted information about Jupiter's composition, temperature, and pressure to the satellite orbiting above. *Galileo* studied Jupiter's moons, rings, and magnetic fields and then relayed this information to scientists who were waiting eagerly for it on Earth.

Studies of Jupiter's moon Europa by *Galileo* indicate that an ocean of water may exist under the surface of Europa. A cracked outer layer of ice makes up Europa's surface, shown in **Figure 11.** The cracks in the surface may be caused by geologic activity that heats the ocean underneath the surface. Sunlight penetrates these cracks, further heating the ocean and setting the stage for the possible existence of life on Europa. *Galileo* ended its study of Europa in 2000. More advanced probes will be needed to determine whether life exists on this icy moon.

Figure 11 Future missions will be needed to determine whether life exists on Europa.

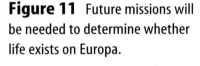 **Reading Check** *What features on Europa suggest the possibility of life existing on this moon?*

In October and November of 1999, *Galileo* approached Io, another one of Jupiter's moons. It came within 300 km and took photographs of a volcanic vent named Loki, which emits more energy than all of Earth's volcanoes combined. *Galileo* also discovered eruption plumes that shoot gas made of sulfur and oxygen.

Moon Quest

Throughout the world, people were shocked when they turned on their radios and television sets in 1957 and heard the radio transmissions from *Sputnik I* as it orbited Earth. All that *Sputnik I* transmitted was a sort of beeping sound, but people quickly realized that launching a human into space wasn't far off.

In 1961, Soviet cosmonaut Yuri A. Gagarin became the first human in space. He orbited Earth and returned safely. Soon, President John F. Kennedy called for the United States to send humans to the Moon and return them safely to Earth. His goal was to achieve this by the end of the 1960s. The race for space was underway.

The U.S. program to reach the Moon began with **Project Mercury.** The goals of Project Mercury were to orbit a piloted spacecraft around Earth and to bring it back safely. The program provided data and experience in the basics of space flight. On May 5, 1961, Alan B. Shepard became the first U.S. citizen in space. In 1962, *Mercury* astronaut John Glenn became the first U.S. citizen to orbit Earth. **Figure 12** shows Glenn preparing for liftoff.

☑ Reading Check *What were the goals of Project Mercury?*

Project Gemini The next step in reaching the Moon was called **Project Gemini.** Teams of two astronauts in the same *Gemini* spacecraft orbited Earth. One *Gemini* team met and connected with another spacecraft in orbit—a skill that would be needed on a voyage to the Moon.

The *Gemini* spacecraft was much like the *Mercury* spacecraft, except it was larger and easier for the astronauts to maintain. It was launched by a rocket known as a *Titan II*, which was a liquid fuel rocket.

In addition to connecting spacecraft in orbit, another goal of Project Gemini was to investigate the effects of space travel on the human body.

Along with the *Mercury* and *Gemini* programs, a series of robotic probes was sent to the Moon. *Ranger* proved that a spacecraft could be sent to the Moon. In 1966, *Surveyor* landed gently on the Moon's surface, indicating that the Moon's surface could support spacecraft and humans. The mission of *Lunar Orbiter* was to take pictures of the Moon's surface that would help determine the best future lunar landing sites.

Figure 12 An important step in the attempt to reach the Moon was John Glenn's first orbit around Earth.

Modeling a Satellite

WARNING: *Stand a safe distance away from classmates.*

Procedure
1. Tie one end of a strong, 50-cm-long **string** to a small **cork.**
2. Hold the other end of the string tightly with your arm fully extended.
3. Move your hand back and forth so that the cork swings in a circular motion.
4. Gradually decrease the speed of the cork.

Analysis
1. What happened as the cork's motion slowed?
2. How does the motion of the cork resemble that of a satellite in orbit?

Project Apollo The final stage of the U.S. program to reach the Moon was **Project Apollo.** On July 20, 1969, *Apollo 11* landed on the Moon's surface. Neil Armstrong was the first human to set foot on the Moon. His first words as he stepped onto its surface were, "That's one small step for man, one giant leap for mankind." Edwin Aldrin, the second of the three *Apollo 11* astronauts, joined Armstrong on the Moon, and they explored its surface for two hours. While they were exploring, Michael Collins remained in the Command Module; Armstrong and Aldrin then returned to the Command Module before beginning the journey home. A total of six lunar landings brought back more than 2,000 samples of moon rock and soil for study before the program ended in 1972. **Figure 13** shows an astronaut exploring the Moon's surface from the Lunar Rover vehicle.

Figure 13 The Lunar Rover vehicle was first used during the *Apollo 15* mission. Riding in the moon buggy, *Apollo 15, 16,* and *17* astronauts explored the lunar surface.

section 2 review

Summary

First Missions into Space

- Rockets are engines that have everything they need to burn fuel.
- Rockets may be fueled with liquid or solid propellants.
- A satellite is any object that revolves around another object.

Space Probes

- A space probe is an instrument that gathers information and sends it back to Earth.
- *Voyager* and *Pioneer* are probes designed to explore the solar system and beyond.
- *Galileo* is a space probe that explored Jupiter and its moons.

Moon Quest

- Project Mercury sent the first piloted spacecraft around Earth.
- *Ranger* and *Surveyor* probes explored the Moon's surface.
- *Gemini* orbited teams of two astronauts.
- Project Apollo completed six lunar landings.

Self Check

1. **Explain** why Neptune has eleven satellites even though it is not orbited by human-made objects.
2. **Explain** why *Galileo* was considered a space probe as it traveled to Jupiter. However, once there, it became an artificial satellite.
3. **List** several discoveries made by the *Voyager 1* and *Voyager 2* space probes.
4. **Sequence** Draw a time line beginning with *Sputnik* and ending with Project Apollo. Include descriptions of important missions.
5. **Think Critically** Is Earth a satellite of any other body in space? Explain.

Applying Math

6. **Solve Simple Equations** A standard unit of measurement in astronomy is the astronomical unit, or AU. It equals is about 150,000,000,000 (1.5×10^{11}) m. In 2000, *Pioneer 10* was more than 11 million km from Earth. How many AUs is this?
7. **Convert Units** A spacecraft is launched at a velocity of 40,200 km/h. Express this speed in kilometers per second. Show your work.

Current and Future Space Missions

The Space Shuttle

Imagine spending millions of dollars to build a machine, sending it off into space, and watching its 3,000 metric tons of metal and other materials burn up after only a few minutes of work. That's exactly what NASA did with the rocket portions of spacecraft for many years. The early rockets were used only to launch a small capsule holding astronauts into orbit. Then sections of the rocket separated from the rest and burned when reentering the atmosphere.

A Reusable Spacecraft NASA administrators, like many others, realized that it would be less expensive and less wasteful to reuse resources. The reusable spacecraft that transports astronauts, satellites, and other materials to and from space is called the **space shuttle**, shown in **Figure 14,** as it is landing.

At launch, the space shuttle stands on end and is connected to an external liquid-fuel tank and two solid-fuel booster rockets. When the shuttle reaches an altitude of about 45 km, the emptied, solid-fuel booster rockets drop off and parachute back to Earth. These are recovered and used again. The external liquid-fuel tank separates and falls back to Earth, but it isn't recovered.

Work on the Shuttle After the space shuttle reaches space, it begins to orbit Earth. There, astronauts perform many different tasks. In the cargo bay, astronauts can conduct scientific experiments and determine the effects of spaceflight on the human body. When the cargo bay isn't used as a laboratory, the shuttle can launch, repair, and retrieve satellites. Then the satellites can be returned to Earth or repaired onboard and returned to space. After a mission, the shuttle glides back to Earth and lands like an airplane. A large landing field is needed as the gliding speed of the shuttle is 335 km/h.

as you read

What You'll Learn

- **Explain** the benefits of the space shuttle.
- **Identify** the usefulness of orbital space stations.
- **Explore** future space missions.
- **Identify** the applications of space technology to everyday life.

Why It's Important

Experiments performed on future space missions may benefit you.

Review Vocabulary
cosmonaut: astronaut of the former Soviet Union or present-day Russian space program

New Vocabulary
- space shuttle
- space station

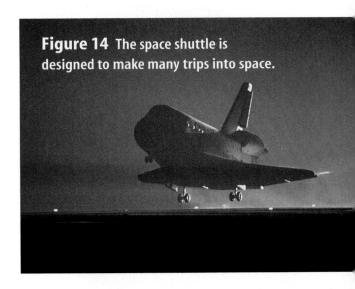

Figure 14 The space shuttle is designed to make many trips into space.

Space Stations

Astronauts can spend only a short time living in the space shuttle. Its living area is small, and the crew needs more room to live, exercise, and work. A **space station** has living quarters, work and exercise areas, and all the equipment and support systems needed for humans to live and work in space.

In 1973, the United States launched the space station *Skylab,* shown in **Figure 15.** Crews of astronauts spent up to 84 days there, performing experiments and collecting data on the effects on humans of living in space. In 1979, the abandoned *Skylab* fell out of orbit and burned up as it entered Earth's atmosphere.

Figure 15 Astronauts performed a variety of tasks while living and working in space onboard *Skylab.*

Crews from the former Soviet Union have spent more time in space, onboard the space station *Mir,* than crews from any other country. Cosmonaut Dr. Valery Polyakov returned to Earth after 438 days in space studying the long-term effects of weightlessness.

Cooperation in Space

In 1995, the United States and Russia began an era of cooperation and trust in exploring space. Early in the year, American Dr. Norman Thagard was launched into orbit aboard the Russian *Soyuz* spacecraft, along with two Russian cosmonaut crewmates. Dr. Thagard was the first U.S. astronaut launched into space by a Russian booster and the first American resident of the Russian space station *Mir.*

Figure 16 Russian and American scientists have worked together to further space exploration.
Explain *why the docking of the space shuttle with* Mir *was so important.*

In June 1995, Russian cosmonauts rode into orbit onboard the space shuttle *Atlantis,* America's 100th crewed launch. The mission of *Atlantis* involved, among other studies, a rendezvous and docking with the space station *Mir.* The cooperation that existed on this mission, as shown in **Figure 16,** continued through eight more space shuttle-*Mir* docking missions. Each of the eight missions was an important step toward building and operating the *International Space Station.* In 2001, the abandoned *Mir* space station fell out of orbit and burned up upon reentering the atmosphere. Cooperation continued as the *International Space Station* began to take form.

The International Space Station The *International Space Station (ISS)* will be a permanent laboratory designed for long-term research projects. Diverse topics will be studied, including research on the growth of protein crystals. This particular project will help scientists determine protein structure and function, which is expected to enhance work on drug design and the treatment of many diseases.

The *ISS* will draw on the resources of 16 nations. These nations will build units for the space station, which then will be transported into space onboard the space shuttle and Russian launch rockets. The station will be constructed in space. **Figure 17** shows what the completed station will look like.

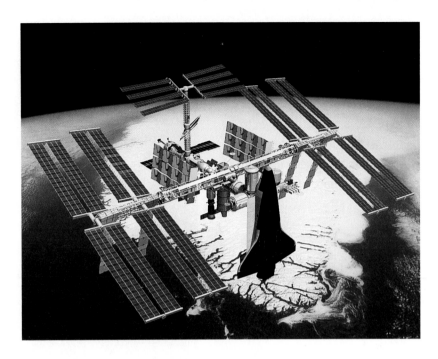

Figure 17 This is a picture of what the proposed *International Space Station* will look like when it is completed in 2006.

 *What is the purpose of the* International Space Station?

Phases of *ISS* NASA is planning the *ISS* program in phases. Phase One, now concluded, involved the space shuttle-*Mir* docking missions. Phase Two began in 1998 with the launch of the Russian-built *Zarya Module,* also known as the Functional Cargo Block. In December 1998, the first assembly of *ISS* occurred when a space shuttle mission attached the Unity module to *Zarya.* During this phase, crews of three people were delivered to the space station. Phase Two ended in 2001 with the addition of a U.S. laboratory.

Living in Space The project will continue with Phase Three when the Japanese Experiment Module, the European Columbus Orbiting Facility, and another Russian lab will be delivered.

It is hoped that the *International Space Station* will be completed in 2006. Eventually, a seven-person crew should be able to work comfortably onboard the station. A total of 47 separate launches will be required to take all the components of the *ISS* into space and prepare it for permanent habitation. NASA plans for crews of astronauts to stay onboard the station for several months at a time. NASA already has conducted numerous tests to prepare crews of astronauts for extended space missions. One day, the station could be a construction site for ships that will travel to the Moon and Mars.

Topic: *International Space Station*

Visit bookj.msscience.com for Web links to information about the *International Space Station.*

Activity You can see the station travel across the sky with an unaided eye. Find out the schedule and try to observe it.

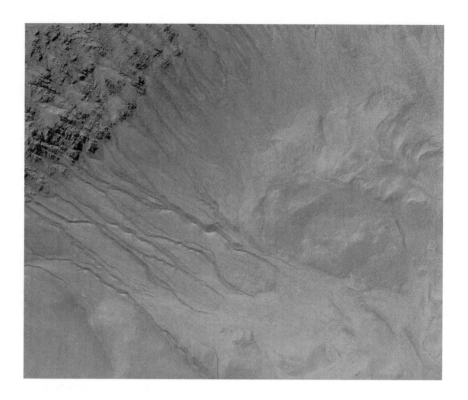

Figure 18 Gulleys, channels, and aprons of sediment imaged by the *Mars Global Surveyor* are similar to features on Earth known to be caused by flowing water. This water is thought to seep out from beneath the surface of Mars.

Exploring Mars

Two of the most successful missions in recent years were the 1996 launchings of the *Mars Global Surveyor* and the *Mars Pathfinder*. *Surveyor* orbited Mars, taking high-quality photos of the planet's surface as shown in **Figure 18.** *Pathfinder* descended to the Martian surface, using rockets and a parachute system to slow its descent. Large balloons absorbed the shock of landing. *Pathfinder* carried technology to study the surface of the planet, including a remote-controlled robot rover called Sojourner. Using information gathered by studying photographs taken by *Surveyor,* scientists determined that water recently had seeped to the surface of Mars in some areas.

✔ **Reading Check** *What type of data were obtained by the* **Mars Global Surveyor?**

Another orbiting spacecraft, the *Mars Odyssey* began mapping the surface of Mars in 2002. Soon after, its data confirmed the findings of *Surveyor*—that Martian soil contains frozen water in the southern polar area. The next step was to send robots to explore the surface of Mars. Twin rovers named *Spirit* and *Opportunity* were launched in 2003 with schedules to reach their separate destinations on Mars in January 2004. Their primary goals are to analyze Martian rocks and soils to tell scientists more about Martian geology and provide clues about the role of water on Mars. Future plans include *Phoenix* in 2008, a robot lander capable of digging over a meter into the surface.

New Millennium Program

To continue space missions into the future, NASA has created the New Millennium Program (NMP). The goal of the NMP is to develop advanced technology that will let NASA send smart spacecraft into the solar system. This will reduce the amount of ground control needed. They also hope to reduce the size of future spacecraft to keep the cost of launching them under control. NASA's challenge is to prove that certain cutting-edge technologies, as well as mission concepts, work in space.

Exploring the Moon

Does water exist in the craters of the Moon's poles? This is one question NASA intends to explore with data gathered from the *Lunar Prospector* spacecraft shown in **Figure 19.** Launched in 1998, the *Lunar Prospector's* one-year mission was to orbit the Moon, mapping its structure and composition. Data obtained from the spacecraft indicate that water ice might be present in the craters at the Moon's poles. Scientists first estimated up to 300 million metric tons of water may be trapped as ice, and later estimates are much higher. In the permanently shadowed areas of some craters, the temperature never exceeds −230°C. Therefore water delivered to the Moon by comets or meteorites early in its history could remain frozen indefinitely.

At the end of its mission, *Lunar Prospector* was deliberately crashed into a lunar crater. Using special telescopes, scientists hoped to see evidence of water vapor thrown up by the collision. None was seen, however scientists still believe that much water ice is there. If so, this water would be useful if a colony is ever built on the Moon.

Topic: New Millennium Program
Visit bookj.msscience.com for Web links to information about NASA's New Millennium Program.

Activity Prepare a table listing proposed missions, projected launch dates, and what they will study.

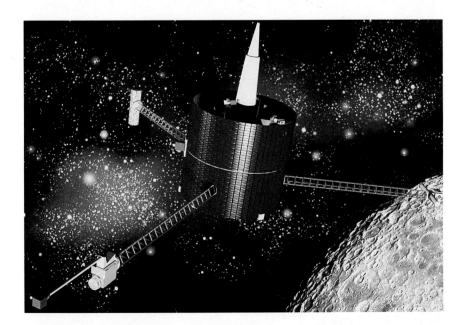

Figure 19 The *Lunar Prospector* analyzed the Moon's composition during its one-year mission. **Explain** *why* Lunar Prospector *was deliberately crashed on the Moon.*

Cassini

In October 1997, NASA launched the space probe *Cassini*. This probe's destination is Saturn. *Cassini*, shown in **Figure 20,** will not reach its goal until 2004. At that time, the space probe will explore Saturn and surrounding areas for four years. One part of its mission is to deliver the European Space Agency's *Huygens* probe to Saturn's largest moon, Titan. Some scientists theorize that Titan's atmosphere may be similar to the atmosphere of early Earth.

Figure 20 *Cassini* is currently on its way to Saturn. After it arrives, it will spend four years studying Saturn and its surrounding area.

The Next Generation Space Telescope Not all space missions involve sending astronauts or probes into space. Plans are being made to launch a new space telescope that is capable of observing the first stars and galaxies in the universe. The *James Webb Space Telescope,* shown in **Figure 21,** will be the successor to the *Hubble Space Telescope.* As part of the Origins project, it will provide scientists with the opportunity to study the evolution of galaxies, the production of elements by stars, and the process of star and planet formation. To accomplish these tasks, the telescope will have to be able to see objects 400 times fainter than those currently studied with ground-based telescopes such as the twin Keck telescopes. NASA hopes to launch the *James Webb Space Telescope* as early as 2010.

Figure 21 The *James Webb Space Telescope* honors the NASA administrator who contributed greatly to the Apollo Program. It will help scientists learn more about how galaxies form.

Everyday Space Technology Many people have benefited from research done for space programs. Medicine especially has gained much from space technology. Space medicine led to better ways to diagnose and treat heart disease here on Earth and to better heart pacemakers. A screening system that works on infants is helping eye doctors spot vision problems early. Cochlear implants that help thousands of deaf people hear were developed using knowledge gained during the space shuttle program.

Space technology can even help catch criminals and prevent accidents. For example, a method to sharpen images that was devised for space studies is being used by police to read numbers on blurry photos of license plates. Equipment using space technology can be placed on emergency vehicles. This equipment automatically changes traffic signals as an emergency vehicle approaches intersections, so that crossing vehicles have time to stop safely. A hand-held device used for travel directions is shown in **Figure 22.**

Figure 22 Global Positioning System (GPS) technology uses satellites to determine location on Earth's surface.

> ✔ **Reading Check** *How have research and technology developed for space benefited people here on Earth?*

---

section 3 review

Summary

The Space Station

- A space station is an orbiting laboratory.
- The new *International Space Station (ISS)* is being built with the aid of 16 nations.
- The space shuttle transports astronauts, satellites, and other materials to and from the *ISS*.

Exploring Mars and the Moon

- The *Mars Global Surveyor* orbited Mars and the *Mars Pathfinder* studied its surface.
- *Lunar Prospector* orbited the Moon, mapping its structure and composition.
- Recent data indicate that water ice crystals may exist in shadows of lunar craters.

Future Missions

- The *Cassini* probe is scheduled to explore Saturn and its moons.
- The successor to the *Hubble* will be the *James Webb Space Telescope*.

Self Check

1. **Identify** the main advantage of the space shuttle.
2. **Describe** the importance of space shuttle-*Mir* docking missions.
3. **Explain** how *International Space Station* is used.
4. **Identify** three ways that space technology is a benefit to everyday life.
5. **Think Critically** What makes the space shuttle more versatile than earlier spacecraft?

Applying Math

6. **Solve One-Step Equations** *Voyager 1* had about 30 kg of hydrazine fuel left in 2003. If it uses about 500 g per year, how long will this fuel last?
7. **Use Percentages** Suppose you're in charge of assembling a crew of 50 people. Decide how many to assign each task, such as farming, maintenance, scientific experimentation, and so on. Calculate the percent of the crew assigned to each task. Justify your decisions.

Use the Internet

Star Sightings

Goals
- **Record** your sightings of Polaris.
- **Share** the data with other students to calculate the circumference of Earth.

Data Source

Science Online

Go to bookj.msscience.com/internet_lab to obtain instructions on how to make an astrolabe. Also visit the Web site for more information about the location of Polaris, and for data from other students.

Safety Precautions

WARNING: *Do not use the astrolabe during the daytime to observe the Sun.*

◉ Real-World Question

For thousands of years, people have measured their position on Earth using the position of Polaris, the North Star. At any given observation point, it always appears at the same angle above the horizon. For example, at the north pole, Polaris appears directly overhead, and at the equator it is just above the northern horizon. Other locations can be determined by measuring the height of Polaris above the horizon using an instrument called an astrolabe. Could you use Polaris to determine the size of Earth? You know that Earth is round. Knowing this, do you think you can estimate the circumference of Earth based on star sightings?

Polaris ———

◉ Make a Plan

1. Obtain an astrolabe or construct one using the instructions posted by visiting the link above.

2. **Design** a data table in your Science Journal similar to the one below.

Polaris Observations

Your Location:

Date	Time	Astrolabe Reading
Do not write in this book.		

3. Decide as a group how you will make your observations. Does it take more than one person to make each observation? When will it be easiest to see Polaris?

▶ Follow Your Plan

1. Make sure your teacher approves your plan before you start.
2. Carry out your observations.
3. **Record** your observations in your data table.
4. Average your readings and post them in the table provided at the link shown below.

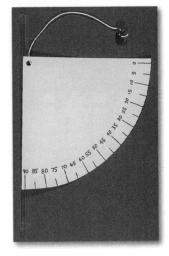

▶ Analyze Your Data

1. **Research** the names of cities that are at approximately the same longitude as your hometown. Gather astrolabe readings from students in one of those cities at the link shown below.
2. **Compare** your astrolabe readings. Subtract the smaller reading from the larger one.
3. **Determine** the distance between your star sighting location and the other city.
4. **Calculate** the circumference of Earth using the following relationship.

   Circumference = (360°) × (distance between locations)/difference between readings

▶ Conclude and Apply

1. **Analyze** how the circumference of Earth that you calculated compares with the accepted value of 40,079 km.
2. **Determine** some possible sources of error in this method of establishing the size of Earth. What improvements would you suggest?

𝒞ommunicating Your Data

Find this lab using the link below. **Create** a poster that includes a table of your data and data from students in other cities. **Perform** a sample circumference calculation for your class.

Scienceonline

bookj.msscience.com/internet_lab

Cities in Space

Should the U.S. spend money to colonize space?

Humans have landed on the Moon, and spacecrafts have landed on Mars. But these space missions are just small steps that may lead to a giant new space program. As technology improves, humans may be able to visit and even live on other planets. But is it worth the time and money involved?

Those in favor of living in space point to the International Space Station that already is orbiting Earth. It's an early step toward establishing floating cities where astronauts can live and work. As Earth's population continues to increase and there is less room on this planet, why not expand to other planets or build a floating city in space? Also, the fact that there is little pollution in space makes the idea appealing to many.

Critics of colonizing space think we should spend the hundreds of billions of dollars that it would cost to colonize space on projects to help improve people's lives here on Earth. Building better housing, developing ways to feed the hungry, finding cures for diseases, and increasing funds for education should come first, these people say. And, critics continue, if people want to explore, why not explore right here on Earth, for example, the ocean floor.

Moon or Mars? If humans were to move permanently to space, the two most likely destinations would be Mars and the Moon, both bleak places. But those in favor of moving to these places say humans could find a way to make them livable as they have made homes in harsh climates and in many rugged areas here on Earth.

Water may be locked in lunar craters, and photos suggest that Mars once had liquid water on its surface. If that water is frozen underground, humans may be able to access it. NASA is studying whether it makes sense to send astronauts and scientists to explore Mars.

Transforming Mars into an Earthlike place with breathable air and usable water will take much longer, but some small steps are being taken. Experimental plants are being developed that could absorb Mars's excess carbon dioxide and release oxygen. Solar mirrors that could warm Mars's surface are available.

Those for and against colonizing space agree on one thing—it will take large amounts of money, research, and planning. It also will take the same spirit of adventure that has led history's pioneers into so many bold frontiers—deserts, the poles, and the sky.

Debate with your class the pros and cons of colonizing space. Do you think the United States should spend money to create space cities or use the money now to improve lives of people on Earth?

Science online

For more information, visit bookj.msscience.com/time

Reviewing Main Ideas

Section 1 Radiation from Space

1. The arrangement of electromagnetic waves according to their wavelengths is the electromagnetic spectrum.

2. Optical telescopes produce magnified images of objects.

3. Radio telescopes collect and record radio waves given off by some space objects.

Section 2 Early Space Missions

1. A satellite is an object that revolves around another object. The moons of planets are natural satellites. Artificial satellites are those made by people.

2. A space probe travels into the solar system, gathers data, and sends them back to Earth.

3. American piloted space programs included the Gemini, Mercury, and Apollo Projects.

Section 3 Current and Future Space Missions

1. Space stations provide the opportunity to conduct research not possible on Earth. The *International Space Station* is being constructed in space with the cooperation of more than a dozen nations.

2. The space shuttle is a reusable spacecraft that carries astronauts, satellites, and other cargo to and from space.

3. Space technology is used to solve problems on Earth, too. Advances in engineering related to space travel have aided medicine, environmental sciences, and other fields.

Visualizing Main Ideas

Copy and complete the following concept map about the race to the Moon. Use the phrases: first satellite, Project Gemini, Project Mercury, team of two astronauts orbits Earth, Project Apollo.

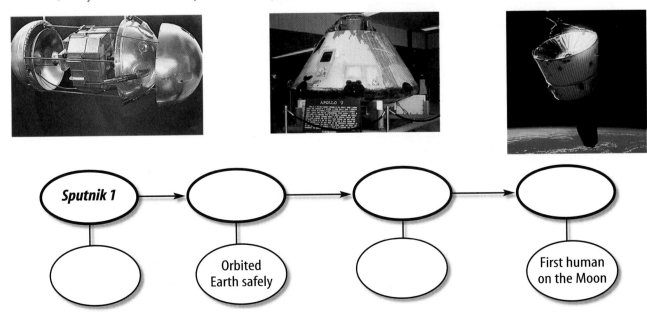

Sputnik 1 → → →

Orbited Earth safely

First human on the Moon

Using Vocabulary

electromagnetic spectrum p. 9	reflecting telescope p. 10
	refracting telescope p. 10
observatory p. 10	rocket p. 15
orbit p. 17	satellite p. 17
Project Apollo p. 22	space probe p. 18
Project Gemini p. 21	space shuttle p. 23
Project Mercury p. 21	space station p. 24
radio telescope p. 13	

Fill in the blanks with the correct vocabulary word(s).

1. A(n) _____ telescope uses lenses to bend light.

2. A(n) _____ is an object that revolves around another object in space.

3. _____ was the first piloted U.S. space program.

4. A(n) _____ carries people and tools to and from space.

5. In the _____, electromagnetic waves are arranged, in order, according to their wavelengths.

Checking Concepts

Choose the word or phrase that best answers the question.

6. Which spacecraft has sent images of Venus to scientists on Earth?
   A) *Voyager*   C) *Apollo 11*
   B) *Viking*   D) *Magellan*

7. Which kind of telescope uses mirrors to collect light?
   A) radio
   B) electromagnetic
   C) refracting
   D) reflecting

8. What was *Sputnik I?*
   A) the first telescope
   B) the first artificial satellite
   C) the first observatory
   D) the first U.S. space probe

9. Which kind of telescope can be used during the day or night and during bad weather?
   A) radio
   B) electromagnetic
   C) refracting
   D) reflecting

10. When fully operational, what is the maximum number of people who will crew the *International Space Station?*
    A) 3   C) 15
    B) 7   D) 50

11. Which space mission's goal was to put a spacecraft into orbit and bring it back safely?
    A) Project Mercury
    B) Project Apollo
    C) Project Gemini
    D) *Viking I*

12. Which of the following is a natural satellite of Earth?
    A) *Skylab*
    B) the space shuttle
    C) the Sun
    D) the Moon

13. What does the space shuttle use to place a satellite into space?
    A) liquid-fuel tank
    B) booster rocket
    C) mechanical arm
    D) cargo bay

14. What part of the space shuttle is reused?
    A) liquid-fuel tanks
    B) *Gemini* rockets
    C) booster engines
    D) Saturn rockets

Science Online bookj.msscience.com/vocabulary_puzzlemaker

Thinking Critically

15. **Compare and contrast** the advantages of a moon-based telescope with an Earth-based telescope.

16. **Infer** how sensors used to detect toxic chemicals in the space shuttle could be beneficial to a factory worker.

17. **Drawing Conclusions** Which do you think is a wiser method of exploration—space missions with people onboard or robotic space probes? Why?

18. **Explain** Suppose two astronauts are outside the space shuttle orbiting Earth. The audio speaker in the helmet of one astronaut quits working. The other astronaut is 1 m away and shouts a message. Can the first astronaut hear the message? Support your reasoning.

19. **Make and Use Tables** Copy and complete the table below. Use information from several resources.

United States Space Probes		
Probe	**Launch Date(s)**	**Planets or Objects Visited**
Vikings 1 and *2*	Do not write in this book.	
Galileo		
Lunar Prospector		
Pathfinder		

20. **Classify** the following as a satellite or a space probe: *Cassini*, *Sputnik I*, *Hubble Space Telescope*, space shuttle, and *Voyager 2*.

21. **Compare and contrast** space probes and artificial satellites.

Performance Activities

22. **Display** Make a display showing some of the images obtained from the *Hubble Space Telescope*. Include samples of three types of galaxies, nebulae, and star clusters.

Applying Math

23. **Space Communications** In May 2003 *Voyager 1* was 13 billion km from the Sun. Calculate how long it takes for a signal to travel this far assuming it travels at 3×10^8 m/s.

Use the graph below to answer question 24.

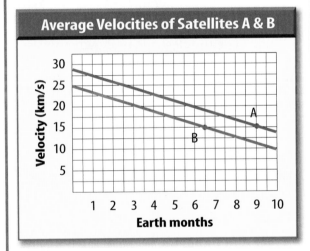

Average Velocities of Satellites A & B

24. **Satellite Orbits** The graph above predicts the average velocities of satellites A and B in orbit around a hypothetical planet. Because of contact with the planet's atmosphere, their velocities are decreasing. At a velocity of 15 km/s their orbits will decay and they will spiral downwards to the surface. Using the graph, determine how long will it take for each satellite to reach this point?

25. **Calculate Fuel** A spacecraft carries 30 kg of hydrazine fuel and uses and average of 500 g/y. How many years could this fuel last?

26. **Space Distances** Find the distance in AUs to a star 68 light-years (LY) distant. (1 LY $=6.3 \times 10^4$ AUs)

Part 1 | Multiple Choice

Record your answers on the answer sheet provided by your teacher or on a sheet of paper.

Use the figure below to answer question 1.

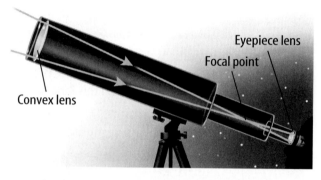

Eyepiece lens

Focal point

Convex lens

1. Which type of telescope is shown above?
   A. refracting   C. reflecting
   B. radio   D. space

2. Who was the first human in space?
   A. Edwin Aldrin   C. Neil Armstrong
   B. John Glenn   D. Yuri Gagarin

3. Which is an engine that can launch an object into space?
   A. space probe   C. capsule
   B. shuttle   D. rocket

4. Which is the speed of light in a vacuum?
   A. 300 km/s   C. 3,000 km/s
   B. 300,000 km/s   D. 30,00 km/s

5. Which of the following is an advantage of space telescopes?
   A. They are cheaper to build.
   B. They have fewer technical problems.
   C. They obtain higher quality images.
   D. They can be repaired easily.

Test-Taking Tip

Making Answers Do not mark the test booklet when taking the test. Be sure to mark ALL answers on your answer sheet and leave no blanks.

6. Which type of radiation has a shorter wavelength than visible light does?
   A. ultraviolet   C. infrared
   B. microwaves   D. radio waves

7. Which space probe visited Mars?
   A. *Viking 1*   C. *Magellan*
   B. *Mariner 2*   D. *Pioneer 10*

8. Which United States space program included several lunar landings?
   A. Gemini   C. Apollo
   B. Mercury   D. space shuttle

Examine the diagram below. Then answer questions 9–11.

9. What is the name of the curved path that the satellite follows?
   A. an orbit   C. a revolution
   B. a rotation   D. a track

10. Which force pulls the satellite toward Earth?
    A. the Moon's gravity
    B. Earth's gravity
    C. the Sun's gravity
    D. Earth's magnetic field

11. Imagine that the satellite in the diagram above started to orbit at a slower speed. Which of the following probably would happen to the satellite?
    A. It would fly off into space.
    B. It would crash into the Moon.
    C. It would crash into the Sun.
    D. It would crash into Earth.

Part 2 | **Short Response/Grid In**

Record your answers on the answer sheet provided by your teacher or on a sheet of paper.

12. Explain the difference between a space probe and a satellite that is orbiting Earth.

13. Why was the flight of *Sputnik 1* important?

14. List four ways that satellites are useful.

15. How are radio telescopes different from optical telescopes?

Use the table below to answer questions 16–19. The table includes data collected by *Mars Pathfinder* on the third Sol, or Martian day, of operation.

Sol 3 Temperature Data from Mars Pathfinder			
Proportion of Sol	**Temperature (°C)**		
	1.0 m above surface	**0.5 m above surface**	**0.25 m above surface**
3.07	−70.4	−70.7	−73.4
3.23	−74.4	−74.9	−75.9
3.33	−53.0	−51.9	−46.7
3.51	−22.3	−19.2	−15.7
3.60	−15.1	−12.5	−8.9
3.70	−26.1	−25.7	−24.0
3.92	−63.9	−64.5	−65.8

16. Which proportion of sol value corresponds to the warmest temperatures at all three heights?

17. Which proportion of sol value corresponds to the coldest temperatures at all three heights?

18. What is the range of the listed temperature values for each distance above the surface?

19. Explain the data in the table. Why do the temperatures vary in this way?

Part 3 | **Open Ended**

Record your answers on a sheet of paper.

Use the diagram below to answer question 20.

Liquid fuel tank

Orbiter

Solid rocket boosters

20. Explain the purpose of each of the labeled objects.

21. List four advancements in technology directly attributable to space exploration and how they have impacted everyday life on Earth.

22. What are the advantages of having reusable spacecraft? Are there any disadvantages? Explain.

23. What is the *International Space Station?* How is it used?

24. What are the advantages of international cooperation during space exploration? Are there disadvantages?

25. Explain how the voices of astronauts onboard the space shuttle can be heard on Earth.

26. List several benefits and costs of space exploration. Do you think that the benefits of space exploration outweigh the costs? Explain why you do or do not.

The Sun-Earth-Moon System

Full Moon Rising— The Real Story

Why does the Moon's appearance change throughout the month? Do the Sun and Moon really rise? You will find the answers to these questions and also learn why we have summer and winter.

Science Journal Rotation or revolution—which motion of Earth brings morning and which brings summer?

Start-Up Activities

Model Rotation and Revolution

The Sun rises in the morning; at least, it seems to. Instead, it is Earth that moves. The movements of Earth cause day and night, as well as the seasons. In this lab, you will explore Earth's movements.

1. Hold a basketball with one finger at the top and one at the bottom. Have a classmate gently spin the ball.

2. Explain how this models Earth's rotation.

3. Continue to hold the basketball and walk one complete circle around another student in your class.

4. How does this model Earth's revolution?

5. **Think Critically** Write a paragraph in your Science Journal describing how these movements of the basketball model Earth's rotation and revolution.

Preview this chapter's content and activities at bookj.msscience.com

Earth and the Moon All on Earth can see and feel the movements of Earth and the Moon as they circle the Sun. Make the following Foldable to organize what you learn about these movements and their effects.

STEP 1 Fold a sheet of paper in half lengthwise.

STEP 2 Fold paper down 2.5 cm from the top. (Hint: From the tip of your index finger to your middle knuckle is about 2.5 cm.)

STEP 3 Open and draw lines along the 2.5-cm fold. Label as shown.

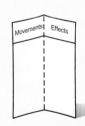

Summarize in a Table As you read the chapter, summarize the movements of Earth and the Moon in the left column and the effects of these movements in the right column.

Earth

as you read

What You'll Learn

- **Examine** Earth's physical characteristics.
- **Differentiate** between rotation and revolution.
- **Discuss** what causes seasons to change.

Why It's Important

Your life follows the rhythm of Earth's movements.

🔍 Review Vocabulary

orbit: the path taken by an object revolving around another

New Vocabulary

- sphere
- ellipse
- axis
- solstice
- rotation
- equinox
- revolution

Figure 1 For many years, sailors have observed that the tops of ships coming across the horizon appear first. This suggests that Earth is spherical, not flat, as was once widely believed.

Properties of Earth

You awaken at daybreak to catch the Sun "rising" from the dark horizon. Then it begins its daily "journey" from east to west across the sky. Finally the Sun "sinks" out of view as night falls. Is the Sun moving—or are you?

It wasn't long ago that people thought Earth was the center of the universe. It was widely believed that the Sun revolved around Earth, which stood still. It is now common knowledge that the Sun only appears to be moving around Earth. Because Earth spins as it revolves around the Sun, it creates the illusion that the Sun is moving across the sky.

Another mistaken idea about Earth concerned its shape. Even as recently as the days of Christopher Columbus, many people believed Earth to be flat. Because of this, they were afraid that if they sailed far enough out to sea, they would fall off the edge of the world. How do you know this isn't true? How have scientists determined the true shape of Earth?

Spherical Shape A round, three-dimensional object is called a **sphere** (SFIHR). Its surface is the same distance from its center at all points. Some common examples of spheres are basketballs and tennis balls.

In the late twentieth century, artificial satellites and space probes sent back pictures showing that Earth is spherical. Much earlier, Aristotle, a Greek astronomer and philosopher who lived around 350 B.C., suspected that Earth was spherical. He observed that Earth cast a curved shadow on the Moon during an eclipse.

In addition to Aristotle, other individuals made observations that indicated Earth's spherical shape. Early sailors, for example, noticed that the tops of approaching ships appeared first on the horizon and the rest appeared gradually, as if they were coming over the crest of a hill, as shown in **Figure 1.**

Additional Evidence Sailors also noticed changes in how the night sky looked. As they sailed north or south, the North Star moved higher or lower in the sky. The best explanation was a spherical Earth.

Today, most people know that Earth is spherical. They also know all objects are attracted by gravity to the center of a spherical Earth. Astronauts have clearly seen the spherical shape of Earth. However, it bulges slightly at the equator and is somewhat flattened at the poles, so it is not a perfect sphere.

Rotation Earth's **axis** is the imaginary vertical line around which Earth spins. This line cuts directly through the center of Earth, as shown in the illustration accompanying **Table 1.** The poles are located at the north and south ends of Earth's axis. The spinning of Earth on its axis, called **rotation,** causes day and night to occur. Here is how it works. As Earth rotates, you can see the Sun come into view at daybreak. Earth continues to spin, making it seem as if the Sun moves across the sky until it sets at night. During night, your area of Earth has rotated so that it is facing away from the Sun. Because of this, the Sun is no longer visible to you. Earth continues to rotate steadily, and eventually the Sun comes into view again the next morning. One complete rotation takes about 24 h, or one day. How many rotations does Earth complete during one year? As you can infer from **Table 1,** it completes about 365 rotations during its one-year journey around the Sun.

Reading Check *Why does the Sun seem to rise and set?*

INTEGRATE Life Science

Earth's Rotation
Suppose that Earth's rotation took twice as long as it does now. In your Science Journal, predict how conditions such as global temperatures, work schedules, plant growth, and other factors might change under these circumstances.

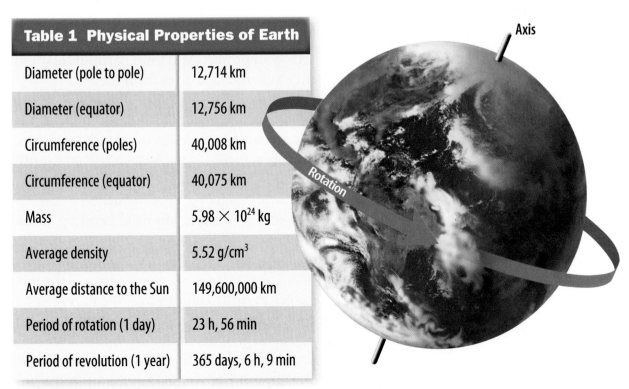

Table 1 Physical Properties of Earth	
Diameter (pole to pole)	12,714 km
Diameter (equator)	12,756 km
Circumference (poles)	40,008 km
Circumference (equator)	40,075 km
Mass	5.98×10^{24} kg
Average density	5.52 g/cm^3
Average distance to the Sun	149,600,000 km
Period of rotation (1 day)	23 h, 56 min
Period of revolution (1 year)	365 days, 6 h, 9 min

Axis

Rotation

Figure 2 Earth's magnetic field is similar to that of a bar magnet, almost as if Earth contained a giant magnet. Earth's magnetic axis is angled 11.5 degrees from its rotational axis.

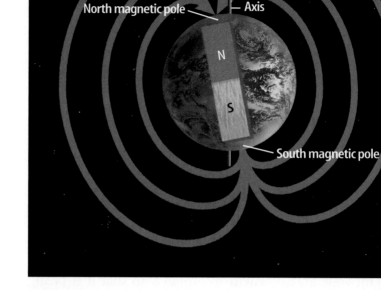

North magnetic pole — Axis

N

S

South magnetic pole

Magnetic Field

Scientists hypothesize that the movement of material inside Earth's core, along with Earth's rotation, generates a magnetic field. This magnetic field is much like that of a bar magnet. Earth has a north and a south magnetic pole, just as a bar magnet has opposite magnetic poles at each of its ends. When you sprinkle iron shavings over a bar magnet, the shavings align with the magnetic field of the magnet. As you can see in **Figure 2,** Earth's magnetic field is similar—almost as if Earth contained a giant bar magnet. Earth's magnetic field protects you from harmful solar radiation by trapping many charged particles from the Sun.

Magnetic Axis When you observe a compass needle pointing north, you are seeing evidence of Earth's magnetic field. Earth's magnetic axis, the line joining its north and south magnetic poles, does not align with its rotational axis. The magnetic axis is inclined at an angle of 11.5° to the rotational axis. If you followed a compass needle, you would end up at the magnetic north pole rather than the rotational north pole.

The location of the magnetic poles has been shown to change slowly over time. The magnetic poles move around the rotational (geographic) poles in an irregular way. This movement can be significant over decades. Many maps include information about the position of the magnetic north pole at the time the map was made. Why would this information be important?

What causes changing seasons?

Flowers bloom as the days get warmer. The Sun appears higher in the sky, and daylight lasts longer. Spring seems like a fresh, new beginning. What causes these wonderful changes?

Orbiting the Sun You learned earlier that Earth's rotation causes day and night. Another important motion is **revolution,** which is Earth's yearly orbit around the Sun. Just as the Moon is Earth's satellite, Earth is a satellite of the Sun. If Earth's orbit were a circle with the Sun at the center, Earth would maintain a constant distance from the Sun. However, this is not the case. Earth's orbit is an **ellipse** (ee LIHPS)—an elongated, closed curve. The Sun is not at the center of the ellipse but is a little toward one end. Because of this, the distance between Earth and the Sun changes during Earth's yearlong orbit. Earth gets closest to the Sun—about 147 million km away—around January 3. The farthest Earth gets from the Sun is about 152 million km away. This happens around July 4 each year.

Reading Check *What is an ellipse?*

Does this elliptical orbit cause seasonal temperatures on Earth? If it did, you would expect the warmest days to be in January. You know this isn't the case in the northern hemisphere, something else must cause the change.

Even though Earth is closest to the Sun in January, the change in distance is small. Earth is exposed to almost the same amount of Sun all year. But the amount of solar energy any one place on Earth receives varies greatly during the year. Next, you will learn why.

A Tilted Axis Earth's axis is tilted 23.5° from a line drawn perpendicular to the plane of its orbit. It is this tilt that causes seasons. The number of daylight hours is greater for the hemisphere, or half of Earth, that is tilted toward the Sun. Think of how early it gets dark in the winter compared to the summer. As shown in **Figure 3,** the hemisphere that is tilted toward the Sun receives more hours of sunlight each day than the hemisphere that is tilted away from the Sun. The longer period of sunlight is one reason summer is warmer than winter, but it is not the only reason.

Science Online

Topic: Ellipses
Visit bookj.msscience.com for Web links to information about orbits and ellipses.

Activity Scientists compare orbits by how close they come to being circular. To do this, they use a measurement called eccentricity. A circle has an eccentricity of zero. Ellipses have eccentricities that are greater than zero, but less than one. The closer the eccentricity is to one, the more elliptical the orbit. Compare the orbits of the four inner planets. List them in order of increasing eccentricity.

Figure 3 In summer, the northern hemisphere is tilted toward the Sun. Notice that the north pole is always lit during the summer. **Observe** *Why is there a greater number of daylight hours in the summer than in the winter?*

North Pole

Radiation from the Sun Earth's tilt also causes the Sun's radiation to strike the hemispheres at different angles. Sunlight strikes the hemisphere tilted towards the Sun at a higher angle, that is, closer to 90 degrees, than the hemisphere tilted away. Thus it receives more total solar radiation than the hemisphere tilted away from the Sun, where sunlight strikes at a lower angle.

Summer occurs in the hemisphere tilted toward the Sun, when its radiation strikes Earth at a higher angle and for longer periods of time. The hemisphere receiving less radiation experiences winter.

Solstices

The **solstice** is the day when the Sun reaches its greatest distance north or south of the equator. In the northern hemisphere, the summer solstice occurs on June 21 or 22, and the winter solstice occurs on December 21 or 22. Both solstices are illustrated in **Figure 4.** In the southern hemisphere, the winter solstice is in June and the summer solstice is in December. Summer solstice is about the longest period of daylight of the year. After this, the number of daylight hours become less and less, until the winter solstice, about the shortest period of daylight of the year. Then the hours of daylight start to increase again.

Figure 4 During the summer solstice in the northern hemisphere, the Sun is directly over the tropic of Cancer, the latitude line at 23.5° N latitude. During the winter solstice, the Sun is directly over the tropic of Capricorn, the latitude line at 23.5° S latitude. At fall and spring equinoxes, the Sun is directly over the equator.

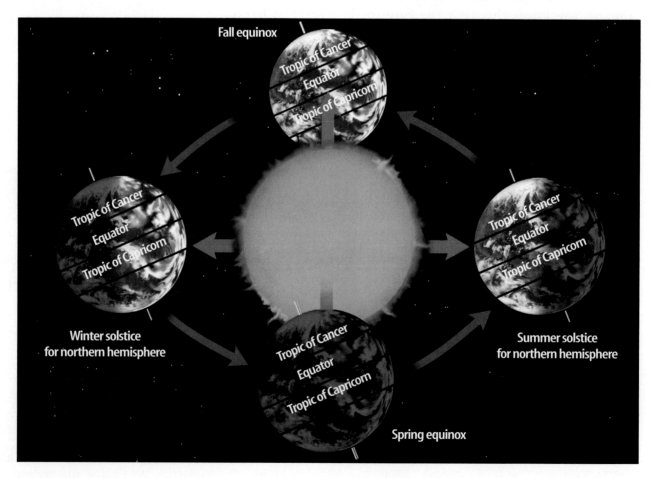

Fall equinox

Tropic of Cancer
Equator
Tropic of Capricorn

Tropic of Cancer
Equator
Tropic of Capricorn

Tropic of Cancer
Equator
Tropic of Capricorn

Tropic of Cancer
Equator
Tropic of Capricorn

**Winter solstice
for northern hemisphere**

**Summer solstice
for northern hemisphere**

Spring equinox

Equinoxes

An **equinox** (EE kwuh nahks) occurs when the Sun is directly above Earth's equator. Because of the tilt of Earth's axis, the Sun's position relative to the equator changes constantly. Most of the time, the Sun is either north or south of the equator, but two times each year it is directly over it, resulting in the spring and fall equinoxes. As you can see in **Figure 4,** at an equinox the Sun strikes the equator at the highest possible angle, 90°.

During an equinox, the number of daylight hours and night-time hours is nearly equal all over the world. Also at this time, neither the northern hemisphere nor the southern hemisphere is tilted toward the Sun.

In the northern hemisphere, the Sun reaches the spring equinox on March 20 or 21, and the fall equinox occurs on September 22 or 23. In the southern hemisphere, the equinoxes are reversed. Spring occurs in September and fall occurs in March.

Earth Data Review As you have learned, Earth is a sphere that rotates on a tilted axis. This rotation causes day and night. Earth's tilted axis and its revolution around the Sun cause the seasons. One Earth revolution takes one year. In the next section, you will read how the Moon rotates on its axis and revolves around Earth.

Science nline

Topic: Seasons
Visit bookj.msscience.com for Web links to information about the seasons.

Activity Make a poster describing how the seasons differ in other parts of the world. Show how holidays might be celebrated differently and how farming might vary between hemispheres.

section 1 review

Summary

Properties of Earth

- Earth is a slightly flattened sphere that rotates around an imaginary line called an axis.
- Earth has a magnetic field, much like a bar magnet.
- The magnetic axis of Earth differs from its rotational axis.

Seasons

- Earth revolves around the Sun in an elliptical orbit.
- The tilt of Earth's axis and its revolution cause the seasons.
- Solstices are days when the Sun reaches its farthest points north or south of the equator.
- Equinoxes are the points when the Sun is directly over the equator.

Self Check

1. **Explain** why Aristotle thought Earth was spherical.
2. **Compare and contrast** rotation and revolution.
3. **Describe** how Earth's distance from the Sun changes throughout the year. When is Earth closest to the Sun?
4. **Explain** why it is summer in Earth's northern hemisphere at the same time it is winter in the southern hemisphere.
5. **Think Critically** **Table 1** lists Earth's distance from the Sun as an average. Why isn't an exact measurement available for this distance?

Applying Skills

6. **Classify** The terms *clockwise* and *counterclockwise* are used to indicate the direction of circular motion. How would you classify the motion of the Moon around Earth as you view it from above Earth's north pole? Now try to classify Earth's movement around the Sun.

The Moon—
Earth's Satellite

as you read

What **You'll Learn**

- **Identify** phases of the Moon and their cause.
- **Explain** why solar and lunar eclipses occur.
- **Infer** what the Moon's surface features may reveal about its history.

Why **It's Important**

Learning about the Moon can teach you about Earth.

Review Vocabulary
mantle: portion of the interior of a planet or moon lying between the crust and core

New Vocabulary
- moon phase
- new moon
- waxing
- full moon
- waning
- solar eclipse
- lunar eclipse
- maria

Motions of the Moon

Just as Earth rotates on its axis and revolves around the Sun, the Moon rotates on its axis and revolves around Earth. The Moon's revolution around Earth is responsible for the changes in its appearance. If the Moon rotates on its axis, why can't you see it spin around in space? The reason is that the Moon's rotation takes 27.3 days—the same amount of time it takes to revolve once around Earth. Because these two motions take the same amount of time, the same side of the Moon always faces Earth, as shown in **Figure 5.**

You can demonstrate this by having a friend hold a ball in front of you. Direct your friend to move the ball in a circle around you while keeping the same side of it facing you. Everyone else in the room will see all sides of the ball. You will see only one side. If the moon didn't rotate, we would see all of its surface during one month.

Figure 5 In about 27.3 days, the Moon orbits Earth. It also completes one rotation on its axis during the same period.
Think Critically *Explain how this affects which side of the Moon faces Earth.*

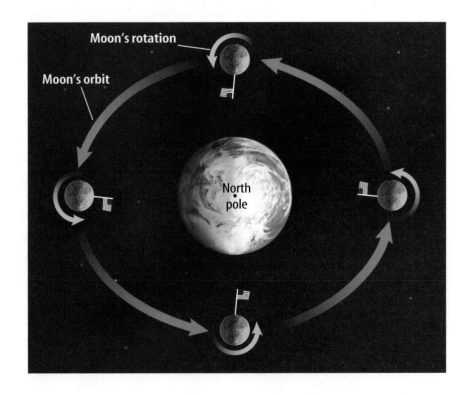

Reflection of the Sun The Moon seems to shine because its surface reflects sunlight. Just as half of Earth experiences day as the other half experiences night, half of the Moon is lighted while the other half is dark. As the Moon revolves around Earth, you see different portions of its lighted side, causing the Moon's appearance to change.

Phases of the Moon

Moon phases are the different forms that the Moon takes in its appearance from Earth. The phase depends on the relative positions of the Moon, Earth, and the Sun, as seen in **Figure 6** on the next page. A **new moon** occurs when the Moon is between Earth and the Sun. During a new moon, the lighted half of the Moon is facing the Sun and the dark side faces Earth. The Moon is in the sky, but it cannot be seen. The new moon rises and sets with the Sun.

Reading Check *Why can't you see a new moon?*

Waxing Phases After a new moon, the phases begin waxing. **Waxing** means that more of the illuminated half of the Moon can be seen each night. About 24 h after a new moon, you can see a thin slice of the Moon. This phase is called the waxing crescent. About a week after a new moon, you can see half of the lighted side of the Moon, or one quarter of the Moon's surface. This is the first quarter phase.

The phases continue to wax. When more than one quarter is visible, it is called waxing *gibbous* after the Latin word for "humpbacked." A **full moon** occurs when all of the Moon's surface facing Earth reflects light.

Waning Phases After a full moon, the phases are said to be waning. When the Moon's phases are **waning,** you see less of its illuminated half each night. Waning gibbous begins just after a full moon. When you can see only half of the lighted side, it is the third-quarter phase. The Moon continues to appear to shrink. Waning crescent occurs just before another new moon. Once again, you can see only a small slice of the Moon.

It takes about 29.5 days for the Moon to complete its cycle of phases. Recall that it takes about 27.3 days for the Moon to revolve around Earth. The discrepancy between these two numbers is due to Earth's revolution. The roughly two extra days are what it takes for the Sun, Earth, and Moon to return to their same relative positions.

Mini LAB

Comparing the Sun and the Moon

Procedure
1. Find an area where you can make a chalk mark on **pavement or similar surface.**
2. Tie a piece of **chalk** to one end of a 200-cm-long **string.**
3. Hold the other end of the string to the pavement.
4. Have a friend pull the string tight and walk around you, drawing a circle (the Sun) on the pavement.
5. Draw a 1-cm-diameter circle in the middle of the larger circle (the Moon).

Analysis
1. How big is the Sun compared to the Moon?
2. The diameter of the Sun is 1.39 million km. The diameter of Earth is 12,756 km. Draw two new circles modeling the sizes of the Sun and Earth. What scale did you use?

Try at Home

Figure 6 The phases of the Moon change during a cycle that lasts about 29.5 days.

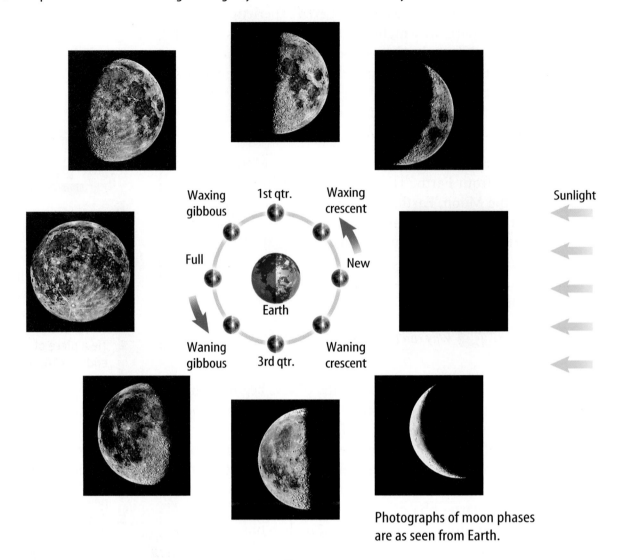

Waxing gibbous

1st qtr.

Waxing crescent

Sunlight

Full

New

Earth

Waning gibbous

3rd qtr.

Waning crescent

Photographs of moon phases are as seen from Earth.

Figure 7 The outer portion of the Sun's atmosphere is visible during a total solar eclipse. It looks like a halo around the Moon.

Eclipses

Imagine living 10,000 years ago. You are foraging for nuts and fruit when unexpectedly the Sun disappears from the sky. The darkness lasts only a short time, and the Sun soon returns to full brightness. You know something strange has happened, but you don't know why. It will be almost 8,000 years before anyone can explain what you just experienced.

The event just described was a total solar eclipse (ih KLIPS), shown in **Figure 7.** Today, most people know what causes such eclipses, but without this knowledge, they would have been terrifying events. During a solar eclipse, many animals act as if it is nighttime. Cows return to their barns and chickens go to sleep. What causes the day to become night and then change back into day?

☑ **Reading Check** *What happens during a total solar eclipse?*

What causes an eclipse? The revolution of the Moon causes eclipses. Eclipses occur when Earth or the Moon temporarily blocks the sunlight from reaching the other. Sometimes, during a new moon, the Moon's shadow falls on Earth and causes a solar eclipse. During a full moon, Earth's shadow can be cast on the Moon, resulting in a lunar eclipse.

An eclipse can occur only when the Sun, the Moon, and Earth are lined up perfectly. Because the Moon's orbit is not in the same plane as Earth's orbit around the Sun, lunar eclipses occur only a few times each year.

Eclipses of the Sun A **solar eclipse** occurs when the Moon moves directly between the Sun and Earth and casts its shadow over part of Earth, as seen in **Figure 8.** Depending on where you are on Earth, you may experience a total eclipse or a partial eclipse. The darkest portion of the Moon's shadow is called the umbra (UM bruh). A person standing within the umbra experiences a total solar eclipse. During a total solar eclipse, the only visible portion of the Sun is a pearly white glow around the edge of the eclipsing Moon.

Surrounding the umbra is a lighter shadow on Earth's surface called the penumbra (puh NUM bruh). Persons standing in the penumbra experience a partial solar eclipse. **WARNING:** *Regardless of which eclipse you view, never look directly at the Sun. The light can permanently damage your eyes.*

Science Online

Topic: Eclipses
Visit bookj.msscience.com for Web links to information about solar and lunar eclipses.

Activity Make a chart showing the dates when lunar and solar eclipses will be visible in your area. Include whether the eclipses will be total or partial.

Figure 8 Only a small area of Earth experiences a total solar eclipse during the eclipse event.

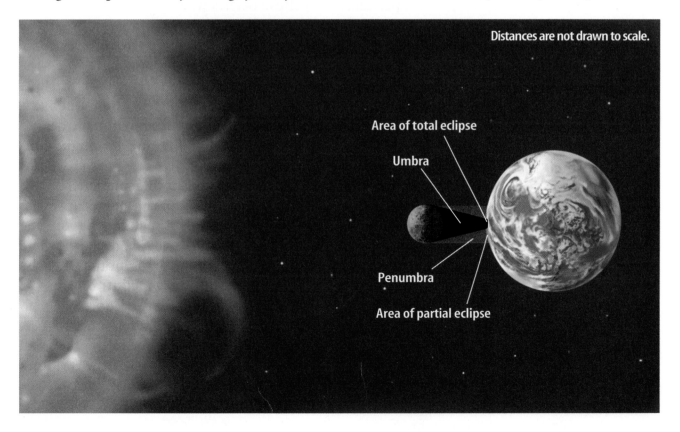

Distances are not drawn to scale.

Area of total eclipse

Umbra

Penumbra

Area of partial eclipse

Figure 9 These photographs show the Moon moving from right to left into Earth's umbra, then out again.

Figure 10 During a total lunar eclipse, Earth's shadow blocks light coming from the Sun.

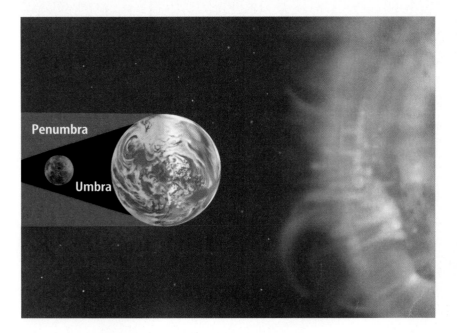

Penumbra

Umbra

Eclipses of the Moon

When Earth's shadow falls on the Moon, a **lunar eclipse** occurs. A lunar eclipse begins when the Moon moves into Earth's penumbra. As the Moon continues to move, it enters Earth's umbra and you see a curved shadow on the Moon's surface, as in **Figure 9.** Upon moving completely into Earth's umbra, as shown in **Figure 10,** the Moon goes dark, signaling that a total lunar eclipse has occurred. Sometimes sunlight bent through Earth's atmosphere causes the eclipsed Moon to appear red.

A partial lunar eclipse occurs when only a portion of the Moon moves into Earth's umbra. The remainder of the Moon is in Earth's penumbra and, therefore, receives some direct sunlight. A penumbral lunar eclipse occurs when the Moon is totally within Earth's penumbra. However, it is difficult to tell when a penumbral lunar eclipse occurs because some sunlight continues to fall on the side of the Moon facing Earth.

A total lunar eclipse can be seen by anyone on the nighttime side of Earth where the Moon is not hidden by clouds. In contrast, only a lucky few people get to witness a total solar eclipse. Only those people in the small region where the Moon's umbra strikes Earth can witness one.

The Moon's Surface

When you look at the Moon, as shown in **Figure 12** on the next page, you can see many depressions called craters. Meteorites, asteroids, and comets striking the Moon's surface created most of these craters, which formed early in the Moon's history. Upon impact, cracks may have formed in the Moon's crust, allowing lava to reach the surface and fill up the large craters. The resulting dark, flat regions are called **maria** (MAHR ee uh). The igneous rocks of the maria are 3 billion to 4 billion years old. So far, they are the youngest rocks to be found on the Moon. This indicates that craters formed after the Moon's surface originally cooled. The maria formed early enough in the Moon's history that molten material still remained in the Moon's interior. The Moon once must have been as geologically active as Earth is today. Before the Moon cooled to the current condition, the interior separated into distinct layers.

Inside the Moon

Earthquakes allow scientists to learn about Earth's interior. In a similar way, scientists use instruments such as the one in **Figure 11** to study moonquakes. The data they have received have led to the construction of several models of the Moon's interior. One such model, shown in **Figure 11,** suggests that the Moon's crust is about 60 km thick on the side facing Earth. On the far side, it is thought to be about 150 km thick. Under the crust, a solid mantle may extend to a depth of 1,000 km. A partly molten zone of the mantle may extend even farther down. Below this mantle may lie a solid, iron-rich core.

INTEGRATE
Career

Seismology A seismologist is an Earth scientist who studies the propagation of seismic waves in geological materials. Usually this means studying earthquakes, but some seismologists apply their knowledge to studies of the Moon and planets. Seismologists usually study geology, physics, and applied mathematics in college and later specialize in seismology for an advanced degree.

Figure 11 Equipment, such as the seismograph left on the Moon by the *Apollo 12* mission, helps scientists study moonquakes.

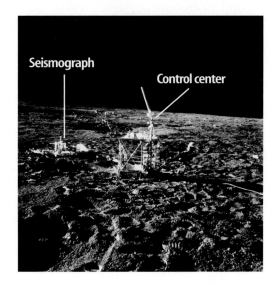

Seismograph

Control center

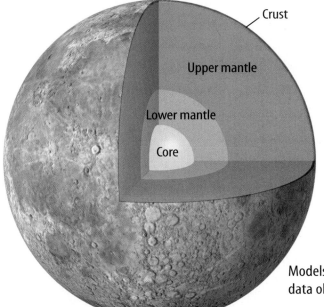

Crust

Upper mantle

Lower mantle

Core

Models of the Moon's interior were created from data obtained by scientists studying moonquakes.

Figure 12

By looking through binoculars, you can see many of the features on the surface of the Moon. These include craters that are hundreds of kilometers wide, light-colored mountains, and darker patches that early astronomers called maria (Latin for "seas"). However, as the NASA Apollo missions discovered, these so-called seas do not contain water. In fact, maria (singular, mare) are flat, dry areas formed by ancient lava flows. Some of the Moon's geographic features are shown below, along with the landing sites of Apollo missions sent to investigate Earth's closest neighbor in space.

NASA astronaut

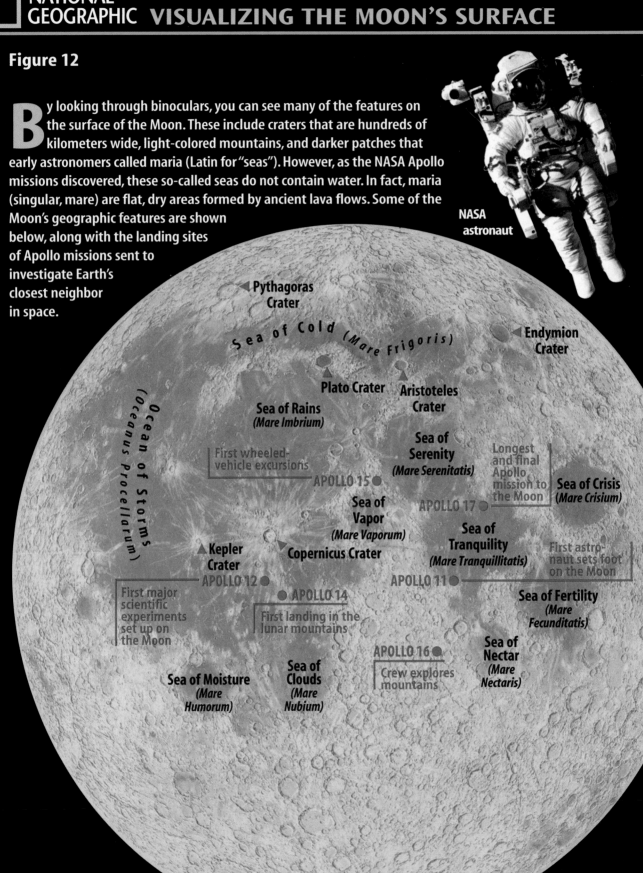

Pythagoras Crater

Sea of Cold (Mare Frigoris)

Endymion Crater

Plato Crater

Aristoteles Crater

Sea of Rains (Mare Imbrium)

Sea of Serenity (Mare Serenitatis)

Ocean of Storms (Oceanus Procellarum)

First wheeled-vehicle excursions

APOLLO 15

Longest and final Apollo mission to the Moon

Sea of Crisis (Mare Crisium)

Sea of Vapor (Mare Vaporum)

APOLLO 17

Kepler Crater

Copernicus Crater

Sea of Tranquility (Mare Tranquillitatis)

First astronaut sets foot on the Moon

APOLLO 12

APOLLO 14

APOLLO 11

First major scientific experiments set up on the Moon

First landing in the lunar mountains

Sea of Fertility (Mare Fecunditatis)

APOLLO 16

Sea of Nectar (Mare Nectaris)

Crew explores mountains

Sea of Moisture (Mare Humorum)

Sea of Clouds (Mare Nubium)

A A Mars-sized object collided with Earth.

B The blast ejected material from both objects into space.

C A ring of gas and debris formed around Earth.

D Particles in the ring joined together to form the Moon.

The Moon's Origin

Before the *Apollo* space missions in the 1960s and 1970s, there were three leading theories about the Moon's origin. According to one theory, the Moon was captured by Earth's gravity. Another held that the Moon and Earth condensed from the same cloud of dust and gas. An alternative theory proposed that Earth ejected molten material that became the Moon.

The Impact Theory The data gathered by the *Apollo* missions have led many scientists to support a new theory, known as the impact theory. It states that the Moon formed billions of years ago from condensing gas and debris thrown off when Earth collided with a Mars-sized object as shown in **Figure 13.**

Figure 13 According to the impact theory, a Mars-sized object collided with Earth around 4.6 billion years ago. Vaporized materials ejected by the collision began orbiting Earth and quickly consolidated into the Moon.

Applying Science

What will you use to survive on the Moon?

You have crash-landed on the Moon. It will take one day to reach a moon colony on foot. The side of the Moon that you are on will be facing away from the Sun during your entire trip. You manage to salvage the following items from your wrecked ship: food, rope, solar-powered heating unit, battery-operated heating unit, oxygen tanks, map of the constellations, compass, matches, water, solar-powered radio transmitter, three flashlights, signal mirror, and binoculars.

Identifying the Problem

The Moon lacks a magnetic field and has no atmosphere. How do the Moon's physical properties and the lack of sunlight affect your decisions?

Solving the Problem

1. Which items will be of no use to you? Which items will you take with you?
2. Describe why each of the salvaged items is useful or not useful.

The Moon in History Studying the Moon's phases and eclipses led to the conclusion that both Earth and the Moon were in motion around the Sun. The curved shadow Earth casts on the Moon indicated to early scientists that Earth was spherical. When Galileo first turned his telescope toward the Moon, he found a surface scarred by craters and maria. Before that time, many people believed that all planetary bodies were perfectly smooth and lacking surface features. Now, actual moon rocks are available for scientists to study, as seen in **Figure 14.** By doing so, they hope to learn more about Earth.

Figure 14 Moon rocks collected by astronauts provide scientists with information about the Moon and Earth.

✔ **Reading Check** *How has observing the Moon been important to science?*

section 2 review

Summary

Motions of the Moon

- The Moon rotates on its axis about once each month.
- The Moon also revolves around Earth about once every 27.3 days.
- The Moon shines because it reflects sunlight.

Phases of the Moon

- During the waxing phases, the illuminated portion of the Moon grows larger.
- During waning phases, the illuminated portion of the Moon grows smaller.
- Earth passing directly between the Sun and the Moon causes a lunar eclipse.
- The Moon passing between Earth and the Sun causes a solar eclipse.

Structure and Origin of the Moon

- The Moon's surface is covered with depressions called impact craters.
- Flat, dark regions within craters are called maria.
- The Moon may have formed as the result of a collision between Earth and a Mars-sized object.

Self Check

1. **Explain** how the Sun, Moon, and Earth are positioned relative to each other during a new moon and how this alignment changes to produce a full moon.

2. **Describe** what phase the Moon must be in to have a lunar eclipse. A solar eclipse?

3. **Define** the terms *umbra* and *penumbra* and explain how they relate to eclipses.

4. **Explain** why lunar eclipses are more common than solar eclipses and why so few people ever have a chance to view a total solar eclipse.

5. **Think Critically** What do the surface features and their distribution on the Moon's surface tell you about its history?

Applying Math

6. **Solve Simple Equations** The Moon travels in its orbit at about 3,400 km/h. Therefore, during a solar eclipse, its shadow sweeps at this speed from west to east. However, Earth rotates from west to east at about 1,670 km/h near the equator. At what speed does the shadow really move across this part of Earth's surface?

M🌑🌑n Phases and E🌙ipses

In this lab, you will demonstrate the positions of the Sun, the Moon, and Earth during certain phases and eclipses. You also will see why only a small portion of the people on Earth witness a total solar eclipse during a particular eclipse event.

◉ Real-World Question

Can a model be devised to show the positions of the Sun, the Moon, and Earth during various phases and eclipses?

Goals
- **Model** moon phases.
- **Model** solar and lunar eclipses.

Materials
light source (unshaded)   globe
polystyrene ball   pencil

Safety Precautions

◉ Procedure

1. Review the illustrations of moon phases and eclipses shown in Section 2.

2. Use the light source as a Sun model and a polystyrene ball on a pencil as a Moon model. Move the Moon around the globe to duplicate the exact position that would have to occur for a lunar eclipse to take place.

3. Move the Moon to the position that would cause a solar eclipse.

4. Place the Moon at each of the following phases: first quarter, full moon, third quarter, and new moon. Identify which, if any, type of eclipse could occur during each phase. Record your data.

Moon Phase Observations	
Moon Phase	**Observations**
First quarter	
Full moon	
Third quarter	Do not write in this book.
New moon	

5. Place the Moon at the location where a lunar eclipse could occur. Move it slightly toward Earth, then away from Earth. Note the amount of change in the size of the shadow.

6. Repeat step 5 with the Moon in a position where a solar eclipse could occur.

◉ Conclude and Apply

1. **Identify** which phase(s) of the Moon make(s) it possible for an eclipse to occur.

2. **Describe** the effect of a small change in distance between Earth and the Moon on the size of the umbra and penumbra.

3. **Infer** why a lunar and a solar eclipse do not occur every month.

4. **Explain** why only a few people have experienced a total solar eclipse.

5. **Diagram** the positions of the Sun, Earth, and the Moon during a first-quarter moon.

6. **Infer** why it might be better to call a full moon a half moon.

Communicate your answers to other students.

Exploring Earth's Moon

as you read

What You'll Learn

- **Describe** recent discoveries about the Moon.
- **Examine** facts about the Moon that might influence future space travel.

Why It's Important

Continuing moon missions may result in discoveries about Earth's origin.

Review Vocabulary

comet: space object orbiting the Sun formed from dust and rock particles mixed with frozen water, methane, and ammonia

New Vocabulary

- impact basin

Missions to the Moon

The Moon has always fascinated humanity. People have made up stories about how it formed. Children's stories even suggested it was made of cheese. Of course, for centuries astronomers also have studied the Moon for clues to its makeup and origin. In 1959, the former Soviet Union launched the first *Luna* spacecraft, enabling up-close study of the Moon. Two years later, the United States began a similar program with the first *Ranger* spacecraft and a series of *Lunar Orbiters*. The spacecraft in these early missions took detailed photographs of the Moon.

The next step was the *Surveyor* spacecraft designed to take more detailed photographs and actually land on the Moon. Five of these spacecraft successfully touched down on the lunar surface and performed the first analysis of lunar soil. The goal of the *Surveyor* program was to prepare for landing astronauts on the Moon. This goal was achieved in 1969 by the astronauts of *Apollo 11*. By 1972, when the *Apollo* missions ended, 12 U.S. astronauts had walked on the Moon. A time line of these important moon missions can be seen in **Figure 15.**

Figure 15 This time line illustrates some of the most important events in the history of moon exploration.

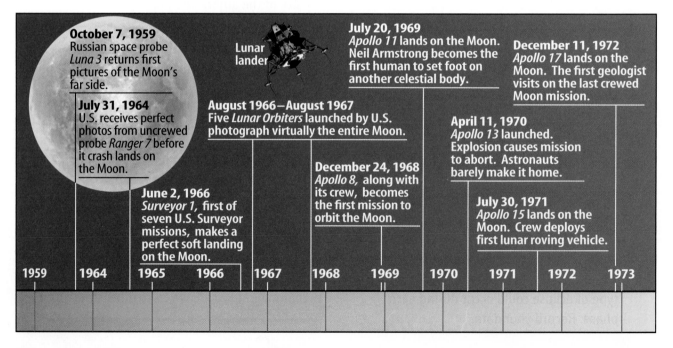

October 7, 1959
Russian space probe *Luna 3* returns first pictures of the Moon's far side.

July 31, 1964
U.S. receives perfect photos from uncrewed probe *Ranger 7* before it crash lands on the Moon.

June 2, 1966
Surveyor 1, first of seven U.S. Surveyor missions, makes a perfect soft landing on the Moon.

Lunar lander

August 1966–August 1967
Five *Lunar Orbiters* launched by U.S. photograph virtually the entire Moon.

December 24, 1968
Apollo 8, along with its crew, becomes the first mission to orbit the Moon.

July 20, 1969
Apollo 11 lands on the Moon. Neil Armstrong becomes the first human to set foot on another celestial body.

April 11, 1970
Apollo 13 launched. Explosion causes mission to abort. Astronauts barely make it home.

July 30, 1971
Apollo 15 lands on the Moon. Crew deploys first lunar roving vehicle.

December 11, 1972
Apollo 17 lands on the Moon. The first geologist visits on the last crewed Moon mission.

| 1959 | 1964 | 1965 | 1966 | 1967 | 1968 | 1969 | 1970 | 1971 | 1972 | 1973 |

Surveying the Moon There is still much to learn about the Moon and, for this reason, the United States resumed its studies. In 1994, the *Clementine* was placed into lunar orbit. Its goal was to conduct a two-month survey of the Moon's surface. An important aspect of this study was collecting data on the mineral content of Moon rocks. In fact, this part of its mission was instrumental in naming the spacecraft. Clementine was the daughter of a miner in the ballad *My Darlin' Clementine*. While in orbit, *Clementine* also mapped features on the Moon's surface, including huge impact basins.

> ✓ **Reading Check** *Why was Clementine placed in lunar orbit?*

Impact Basins When meteorites and other objects strike the Moon, they leave behind depressions in the Moon's surface. The depression left behind by an object striking the Moon is known as an **impact basin,** or impact crater. The South Pole-Aitken Basin is the oldest identifiable impact feature on the Moon's surface. At 12 km in depth and 2,500 km in diameter, it is also the largest and deepest impact basin in the solar system.

Impact basins at the poles were of special interest to scientists. Because the Sun's rays never strike directly, the crater bottoms remain always in shadow. Temperatures in shadowed areas, as shown in **Figure 16,** would be extremely low, probably never more than −173°C. Scientists hypothesize that any ice deposited by comets impacting the Moon throughout its history would remain in these shadowed areas. Indeed, early signals from *Clementine* indicated the presence of water. This was intriguing, because it could be a source of water for future moon colonies.

Science Online

Topic: The Far Side
Visit bookj.msscience.com for Web links to information about the far side of the Moon.

Activity Compare the image of the far side of the Moon with that of the near side shown in **Figure 12.** Make a list of all the differences you note and then compare them with lists made by other students.

Figure 16 The South Pole-Aitken Basin is the largest of its kind found anywhere in the solar system. The deepest craters in the basin stay in shadow throughout the Moon's rotation. Ice deposits from impacting comets are thought to have collected at the bottom of these craters.

Figure 17 This computer-enhanced map based on *Clementine* data indicates the thickness of the Moon's crust. The crust of the side of the Moon facing Earth, shown mostly in red, is thinner than the crust on the far side of the Moon.

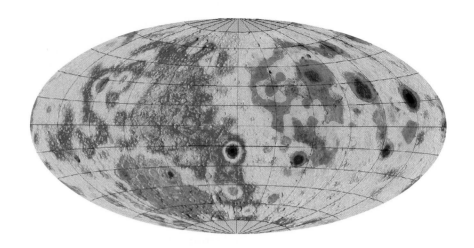

Mapping the Moon

A large part of *Clementine's* mission included taking high-resolution photographs so a detailed map of the Moon's surface could be compiled. *Clementine* carried cameras and other instruments to collect data at wavelengths ranging from infrared to ultraviolet. One camera could resolve features as small as 20 m across. One image resulting from *Clementine* data is shown in **Figure 17.** It shows that the crust on the side of the Moon that faces Earth is much thinner than the crust on the far side. Additional information shows that the Moon's crust is thinnest under impact basins. Based on analysis of the light data received from *Clementine,* a global map of the Moon also was created that shows its composition, as seen in **Figure 18.**

✓ **Reading Check** *What information about the Moon did scientists learn from* Clementine*?*

The Lunar Prospector The success of Clementine opened the door for further moon missions. In 1998, NASA launched the desk-sized *Lunar Prospector,* shown in **Figure 18,** into lunar orbit. The spacecraft spent a year orbiting the Moon from pole to pole, once every two hours. The resulting maps confirmed the *Clementine* data. Also, data from *Lunar Prospector* confirmed that the Moon has a small, iron-rich core about 600 km in diameter. A small core supports the impact theory of how the Moon formed—only a small amount of iron could be blasted away from Earth.

Figure 18 *Lunar Prospector* performed high-resolution mapping of the lunar surface and had instruments that detected water ice at the lunar poles.

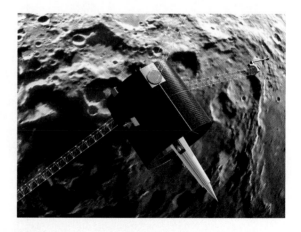

Icy Poles In addition to photographing the surface, *Lunar Prospector* carried instruments designed to map the Moon's gravity, magnetic field, and the abundances of 11 elements in the lunar crust. This provided scientists with data from the entire lunar surface rather than just the areas around the Moon's equator, which had been gathered earlier. Also, *Lunar Prospector* confirmed the findings of *Clementine* that water ice was present in deep craters at both lunar poles.

Later estimates concluded that as much as 3 billion metric tons of water ice was present at the poles, with a bit more at the north pole. Using data from *Lunar Prospector,* scientists prepared maps showing the location of water ice at each pole. **Figure 19** shows how water may be distributed at the Moon's north pole. At first it was thought that ice crystals were mixed with lunar soil, but most recent results suggest that the ice may be in the form of more compact deposits.

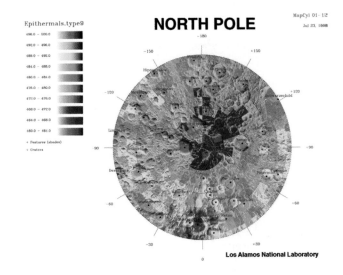

Figure 19 The *Lunar Prospector* data indicates that ice exists in crater shadows at the Moon's poles.

---

section 3 review

Summary

Missions to the Moon

- The first lunar surveys were done by *Luna,* launched by the former Soviet Union, and U.S.-launched *Ranger* and *Lunar Orbiters.*
- Five *Surveyor* probes landed on the Moon.
- U.S. Astronauts landed on and explored the Moon in the *Apollo* program.
- *Clementine,* a lunar orbiter, mapped the lunar surface and collected data on rocks.
- *Clementine* found that the lunar crust is thinner on the side facing Earth.
- Data from *Clementine* indicated that water ice could exist in shaded areas of impact basins.

Mapping the Moon

- *Lunar Prospector* orbited the Moon from pole to pole, collecting data that confirm *Clementine* results and that the Moon has a small iron-rich core.
- Data from *Lunar Prospector* indicate the presence of large quantities of water ice in craters at the lunar poles.

Self Check

1. **Name** the first U.S. spacecraft to successfully land on the Moon. What was the major purpose of this program?

2. **Explain** why scientists continue to study the Moon long after the *Apollo* program ended and list some of the types of data that have been collected.

3. **Explain** how water ice might be preserved in portions of deep impact craters.

4. **Describe** how the detection of a small iron-rich core supports the theory that the Moon was formed from a collision between Earth and a Mars-sized object.

5. **Think Critically** Why might the discovery of ice in impact basins at the Moon's poles be important to future space flights?

Applying Skills

6. **Infer** why it might be better to build a future moon base on a brightly lit plateau near a lunar pole in the vicinity of a deep crater. Why not build a base in the crater itself?

TILT AND TEMPERATURE

Goals

- **Measure** the temperature change in a surface after light strikes it at different angles.
- **Describe** how the angle of light relates to seasons on Earth.

Materials

tape
black construction paper
   (one sheet)
gooseneck lamp
   with 75-watt bulb
Celsius thermometer
watch
protractor

Safety Precautions

WARNING: *Do not touch the lamp without safety gloves. The lightbulb and shade can be hot even when the lamp has been turned off. Handle the thermometer carefully. If it breaks, do not touch anything. Inform your teacher immediately.*

If you walk on blacktop pavement at noon, you can feel the effect of solar energy. The Sun's rays hit at the highest angle at midday. Now consider the fact that Earth is tilted on its axis. How does this tilt affect the angle at which light rays strike an area on Earth? How is the angle of the light rays related to the amount of heat energy and the changing seasons?

⊙ Real-World Question

How does the angle at which light strikes Earth affect the amount of heat energy received by any area on Earth?

⊙ Procedure

1. Choose three angles that you will use to aim the light at the paper.
2. **Determine** how long you will shine the light at each angle before you measure the temperature. You will measure the temperature at two times for each angle. Use the same time periods for each angle.
3. Copy the following data table into your Science Journal to record the temperature the paper reaches at each angle and time.

Temperature Data

Angle of Lamp	Initial Temperature	Temperature at ____ Minutes/Seconds	Temperature at ____ Minutes/Seconds
First angle			
Second angle		Do not write in this book.	
Third angle			

4. Form a pocket out of a sheet of black construction paper and tape it to a desk or the floor.
5. Using the protractor, set the gooseneck lamp so that it will shine on the paper at one of the angles you chose.

6. Place the thermometer in the paper pocket. Turn on the lamp. Use the thermometer to measure the temperature of the paper at the end of the first time period. Continue shining the lamp on the paper until the second time period has passed. Measure the temperature again. Record your data in your data table.

7. Turn off the lamp until the paper cools to room temperature. Repeat steps 5 and 6 using your other two angles.

◉ Conclude and Apply

1. **Describe** your experiment. Identify the variables in your experiment. Which were your independent and dependent variables?

2. **Graph** your data using a line graph. Describe what your graph tells you about the data.

3. **Describe** what happened to the temperature of the paper as you changed the angle of light.

4. **Predict** how your results might have been different if you used white paper. Explain why.

5. **Describe** how the results of this experiment apply to seasons on Earth.

𝒞ommunicating Your Data

Compare your results with those of other students in your class. **Discuss** how the different angles and time periods affected the temperatures.

THE Mayan Calendar

Most people take for granted that a week is seven days, and that a year is 12 months. However there are other ways to divide time into useful units. Roughly 1,750 years ago, in what is now south Mexico and Central America, the Mayan people invented a calendar system based on careful observations of sun and moon cycles.

These glyphs represent four different days of the Tzolkin calendar.

In fact, the Maya had several calendars that they used at the same time.

Two calendars were most important—one was based on 260 days and the other on 365 days. The calendars were so accurate and useful that later civilizations, including the Aztecs, adopted them.

The 260-Day Calendar

This Mayan calendar, called the *Tzolkin* (tz uhl KIN), was used primarily to time planting, harvesting, drying, and storing of corn—their main crop. Each day of the *Tzolkin* had one of 20 names, as well as a number from 1 to 13 and a Mayan god associated with it.

The 365-Day Calendar

Another Mayan calendar, called the *Haab* (HAHB), was based on the orbit of Earth around the Sun. It was divided into 18 months with 20 days each, plus five extra days at the end of each year.

These calendars were used together making the Maya the most accurate reckoners of time before the modern period. In fact, they were only one day off every 6,000 years.

The Kukulkan, built around the year 1050 A.D., in what is now Chichén Itzá, Mexico, was used by the Maya as a calendar. It had four stairways, each with 91 steps, a total of 365 including the platform on top.

Drawing Symbols The Maya created picture symbols for each day of their week. Historians call these symbols glyphs. Collaborate with another student to invent seven glyphs—one for each weekday. Compare them with other glyphs at msscience.com/time.

Science online

For more information, visit bookj.msscience.com/time

Reviewing Main Ideas

Section 1 Earth

1. Earth is spherical and bulges slightly at its equator.

2. Earth rotates once per day and orbits the Sun in a little more than 365 days.

3. Earth has a magnetic field.

4. Seasons on Earth are caused by the tilt of Earth's axis as it orbits the Sun.

Section 2 The Moon— Earth's Satellite

1. Earth's Moon goes through phases that depend on the relative positions of the Sun, the Moon, and Earth.

2. Eclipses occur when Earth or the Moon temporarily blocks sunlight from reaching the other.

3. The Moon's maria are the result of ancient lava flows. Craters on the Moon's surface formed from impacts with meteorites, asteroids, and comets.

Section 3 Exploring Earth's Moon

1. The *Clementine* spacecraft took detailed photographs of the Moon's surface and collected data indicating the presence of water in deep craters.

2. NASA's *Lunar Prospector* spacecraft found additional evidence of ice.

Visualizing Main Ideas

Copy and complete the following concept map on the impact theory of the Moon's formation.

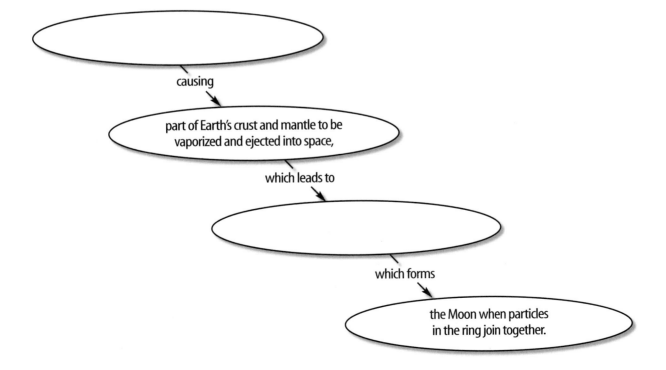

causing

part of Earth's crust and mantle to be vaporized and ejected into space,

which leads to

which forms

the Moon when particles in the ring join together.

Using Vocabulary

axis p. 41	new moon p. 47
ellipse p. 43	revolution p. 43
equinox p. 45	rotation p. 41
full moon p. 47	solar eclipse p. 49
impact basin p. 57	solstice p. 44
lunar eclipse p. 50	sphere p. 40
maria p. 51	waning p. 47
moon phase p. 47	waxing p. 47

Fill in the blanks with the correct vocabulary word or words.

1. The spinning of Earth around its axis is called _____.

2. The _____ is the point at which the Sun reaches its greatest distance north or south of the equator.

3. The Moon is said to be _____ when less and less of the side facing Earth is lighted.

4. The depression left behind by an object striking the Moon is called a(n) _____.

5. Earth's orbit is a(n) _____.

Checking Concepts

Choose the word or phrase that best answers the question.

6. How long does it take for the Moon to rotate once?
   A) 24 hours   C) 27.3 hours
   B) 365 days   D) 27.3 days

7. Where is Earth's circumference greatest?
   A) equator   C) poles
   B) mantle   D) axis

8. Earth is closest to the Sun during which season in the northern hemisphere?
   A) spring   C) winter
   B) summer   D) fall

9. What causes the Sun to appear to rise and set?
   A) Earth's revolution
   B) the Sun's revolution
   C) Earth's rotation
   D) the Sun's rotation

Use the photo below to answer question 10.

10. What phase of the Moon is shown in the photo above?
    A) waning crescent   C) third quarter
    B) waxing gibbous   D) waning gibbous

11. How long does it take for the Moon to revolve once around Earth?
    A) 24 hours   C) 27.3 hours
    B) 365 days   D) 27.3 days

12. What is it called when the phases of the Moon appear to get larger?
    A) waning   C) rotating
    B) waxing   D) revolving

13. What kind of eclipse occurs when the Moon blocks sunlight from reaching Earth?
    A) solar   C) full
    B) new   D) lunar

14. What is the darkest part of the shadow during an eclipse?
    A) waxing gibbous   C) waning gibbous
    B) umbra   D) penumbra

15. What is the name for a depression on the Moon caused by an object striking its surface?
    A) eclipse   C) phase
    B) moonquake   D) impact basin

Thinking Critically

16. **Predict** how the Moon would appear to an observer in space during its revolution. Would phases be observable? Explain.

17. **Predict** what the effect would be on Earth's seasons if the axis were tilted at 28.5° instead of 23.5°.

18. **Infer** Seasons in the two hemispheres are opposite. Explain how this supports the statement that seasons are NOT caused by Earth's changing distance from the Sun.

19. **Draw Conclusions** How would solar eclipses be different if the Moon were twice as far from Earth? Explain.

20. **Predict** how the information gathered by moon missions could be helpful in the future for people wanting to establish a colony on the Moon.

21. **Use Variables, Constants, and Controls** Describe a simple activity to show how the Moon's rotation and revolution work to keep the same side facing Earth at all times.

22. **Draw Conclusions** Gravity is weaker on the Moon than it is on Earth. Why might more craters be present on the far side of the Moon than on the side of the Moon facing Earth?

23. **Recognize Cause and Effect** During a new phase of the Moon, we cannot see it because no sunlight reaches the side facing Earth. Yet sometimes when there is a thin crescent visible, we do see a faint image of the rest of the Moon. Explain what might cause this to happen.

24. **Describe** Earth's magnetic field. Include an explanation of how scientists believe it is generated and two ways in which it helps people on Earth.

Performance Activities

25. **Display** Draw a cross section of the Moon. Include the crust, outer and inner mantles, and possible core based on the information in this chapter. Indicate the presence of impact craters and show how the thickness of the crust varies from one side of the Moon to the other.

26. **Poem** Write a poem in which you describe the various surface features of the Moon. Be sure to include information on how these features formed.

Applying Math

27. **Orbital Tilt** The Moon's orbit is tilted at an angle of 5° to Earth's orbit around the sun. Using a protractor, draw the Moon's orbit around Earth. What fraction of a full circle (360°) is 5°?

Use the illustration below to answer question 28.

28. **Model to Scale** You are planning to make a scale model of the *Lunar Prospector* spacecraft, shown above. Assuming that the three instrument masts are of equal length, draw a labeled diagram of your model using a scale of 1 cm equals 30 cm.

29. **Spacecraft Velocity** The *Lunar Prospector* spacecraft shown above took 105 hours to reach the Moon. Assuming that the average distance from Earth to Moon is 384,000 km, calculate its average velocity on the trip.

Part 1 Multiple Choice

Record your answers on the answer sheet provided by your teacher or on a sheet of paper.

1. Which of the following terms would you use to describe the spinning of Earth on its axis?
   A. revolution
   B. ellipse
   C. rotation
   D. solstice

Use the illustration below to answer questions 2 and 3.

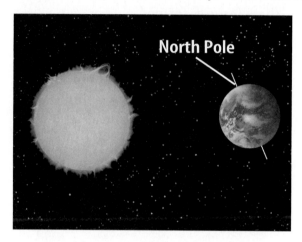

North Pole

2. Which season is beginning for the southern hemisphere when Earth is in this position?
   A. spring
   B. summer
   C. fall
   D. winter

3. Which part of Earth receives the greatest total amount of solar radiation when Earth is in this position?
   A. northern hemisphere
   B. South Pole
   C. southern hemisphere
   D. equator

4. Which term describes the dark, flat areas on the Moon's surface which are made of cooled, hardened lava?
   A. spheres
   B. moonquakes
   C. highlands
   D. maria

Use the illustration below to answer questions 5 and 6.

— Time —→

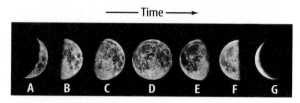

A   B   C   D   E   F   G

5. Which letter corresponds to the moon phase waning gibbous?
   A. G
   B. C
   C. E
   D. A

6. The Moon phase cycle lasts about 29.5 days. Given this information, about how long does it take the Moon to wax from new moon to full moon?
   A. about 3 days
   B. about 1 week
   C. about 2 weeks
   D. about 4 weeks

7. Where have large amounts of water been detected on the Moon?
   A. highlands
   B. lunar equator
   C. maria
   D. lunar poles

8. In what month is Earth closest to the Sun?
   A. March
   B. September
   C. July
   D. January

9. So far, where on the Moon have the youngest rocks been found?
   A. lunar highlands
   B. maria
   C. lunar poles
   D. lunar equator

Test-Taking Tip

Eliminate Choices If you don't know the answer to a multiple-choice question, eliminate as many incorrect choices as possible. Mark your best guess from the remaining answers before moving to the next question.

Question 5 Eliminate those phases that you know are not gibbous.

Part 2 | Short Response/Grid In

Record your answers on the answer sheet provided by your teacher or on a sheet of paper.

10. Explain why the North Pole is always in sunlight during summer in the northern hemisphere.

11. Describe one positive effect of Earth's magnetic field.

12. Explain the difference between a solstice and an equinox. Give the dates of these events on Earth.

13. Explain how scientists know that the Moon was once geologically active.

Use the illustration below to answer questions 14 and 15.

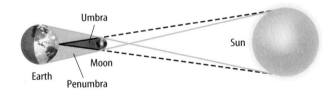

14. What type of eclipse is shown above?

15. Describe what a person standing in the Moon's umbra would see if he or she looked at the sky wearing protective eyewear.

16. The tilt of Earth on its axis causes seasons. Give two reasons why this tilt causes summer to be warmer than winter.

17. When the Apollo missions ended in 1972, 12 astronauts had visited the Moon and brought back samples of moon soil and rock. Explain why we continue to send orbiting spacecraft to study the Moon.

18. Define the term *impact basin*, and name the largest one known in the solar system.

Part 3 | Open Ended

Record your answers on a sheet of paper.

Use the illustrations below to answer questions 19 and 20.

19. As a ship comes into view over the horizon, the top appears before the rest of the ship. How does this demonstrate that Earth is spherical?

20. If Earth were flat, how would an approaching ship appear differently?

21. Explain why eclipses of the Sun occur only occasionally despite the fact that the Moon's rotation causes it to pass between Earth and the Sun every month.

22. Recent data from the spacecraft *Lunar Prospector* indicate the presence of large quantities of water in shadowed areas of lunar impact basins. Describe the hypothesis that scientists have developed to explain how this water reached the moon and how it might be preserved.

23. Compare the impact theory of lunar formation with one of the older theories proposed before the *Apollo* mission.

24. Describe how scientists study the interior of the Moon and what they have learned so far.

25. Explain why Earth's magnetic north poles must be mapped and why these maps must be kept up-to-date.

The Solar System

How is space explored?

You've seen the Sun and the Moon. You also might have observed some of the planets. But to get a really good look at the solar system from Earth, telescopes are needed. The optical telescope at the Keck Observatory in Hawaii allows scientists a close-up view.

Science Journal If you could command the Keck telescope, what would you view? Describe what you would see.

Start-Up Activities

Model Crater Formation

Some objects in the solar system have many craters. The Moon is covered with them. The planet Mercury also has a cratered landscape. Even Earth has some craters. All of these craters formed when rocks from space hit the surface of the planet or moon. In this lab, you'll explore crater formation.

1. Place white flour into a metal cake pan to a depth of 3 cm.

2. Cover the flour with 1 cm of colored powdered drink mix or different colors of gelatin powder.

3. From different heights, ranging from 10 cm to 25 cm, drop various-sized marbles into the pan.

4. **Think Critically** Make drawings in your Science Journal that show what happened to the surface of the powder when marbles were dropped from different heights.

Preview this chapter's content and activities at bookj.msscience.com

The Solar System Make the following Foldable to help you identify what you already know, what you want to know, and what you learned about the solar system.

STEP 1 **Fold** a vertical sheet of paper from side to side. Make the front edge about 1.25 cm shorter than the back edge.

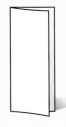

STEP 2 **Turn** lengthwise and **fold** into thirds.

STEP 3 **Unfold and cut** only the top layer along both folds to make three tabs.

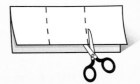

STEP 4 **Label** each tab.

Identify Questions Before you read the chapter, write what you already know about the solar system under the left tab of your Foldable. Write questions about what you'd like to know under the center tab. After you read the chapter, list what you learned under the right tab.

The Solar System

as you read

What You'll Learn

- **Compare** models of the solar system.
- **Explain** that gravity holds planets in orbits around the Sun.

Why It's Important

New technology has come from exploring the solar system.

Review Vocabulary

system: a portion of the universe and all of its components, processes, and interactions

New Vocabulary

- solar system

Ideas About the Solar System

People have been looking at the night sky for thousands of years. Early observers noted the changing positions of the planets and developed ideas about the solar system based on their observations and beliefs. Today, people know that objects in the solar system orbit the Sun. People also know that the Sun's gravity holds the solar system together, just as Earth's gravity holds the Moon in its orbit around Earth. This wasn't always accepted as fact.

Earth-Centered Model Many early Greek scientists thought the planets, the Sun, and the Moon were fixed in separate spheres that rotated around Earth. The stars were thought to be in another sphere that also rotated around Earth. This is called the Earth-centered model of the solar system. It included Earth, the Moon, the Sun, five planets—Mercury, Venus, Mars, Jupiter, and Saturn—and the sphere of stars.

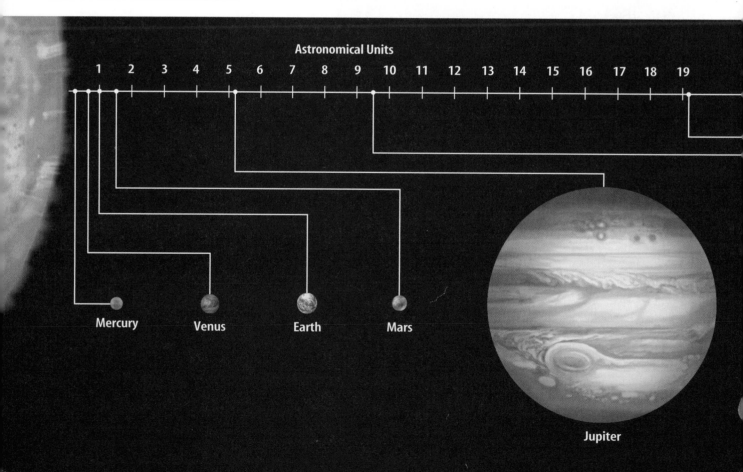

Astronomical Units

1 2 3 4 5 6 7 8 9 10 11 12 13 14 15 16 17 18 19

Mercury   Venus   Earth   Mars

Jupiter

Sun-Centered Model People believed the idea of an Earth-centered solar system for centuries. Then in 1543, Nicholas Copernicus published a different view. Copernicus stated that the Moon revolved around Earth and that Earth and the other planets revolved around the Sun. He also stated that the daily movement of the planets and the stars was caused by Earth's rotation. This is the Sun-centered model of the solar system.

Using his telescope, Galileo Galilei observed that Venus went through a full cycle of phases like the Moon's. He also observed that the apparent diameter of Venus was smallest when the phase was near full. This only could be explained if Venus were orbiting the Sun. Galileo concluded that the Sun is the center of the solar system.

Modern View of the Solar System We now know that the **solar system** is made up of nine planets, including Earth, and many smaller objects that orbit the Sun. The nine planets and the Sun are shown in **Figure 1.** Notice how small Earth is compared with some of the other planets and the Sun.

The solar system includes a huge volume of space that stretches in all directions from the Sun. Because the Sun contains 99.86 percent of the mass of the solar system, its gravity is immense. The Sun's gravity holds the planets and other objects in the solar system in their orbits.

Topic: Solar System
Visit bookj.msscience.com for Web links to information about the solar system.

Activity Make a list of objects in the solar system. Write a one-sentence description of each object on your list.

Figure 1 Each of the nine planets in the solar system is unique. The distances between the planets and the Sun are shown on the scale. One astronomical unit (AU) is the average distance between Earth and the Sun.

Rotational Motion You might have noticed that when a twirling ice skater pulls in her arms, she spins faster. The same thing occurs when a cloud of gas, ice, and dust in a nebula contracts. As mass moves toward the center of the cloud, the cloud rotates faster.

How the Solar System Formed

Scientists hypothesize that the solar system formed from part of a nebula of gas, ice, and dust, like the one shown in **Figure 2.** Follow the steps shown in **Figures 3A** through **3D** to learn how this might have happened. A nearby star might have exploded or nearby O- or B-type stars formed, and the shock waves produced by these events could have caused the cloud to start contracting. As it contracted, the nebula likely fragmented into smaller and smaller pieces. The density in the cloud fragments became greater, and the attraction of gravity pulled more gas and dust toward several centers of contraction. This in turn caused them to flatten into disks with dense centers. As the cloud fragments continued to contract, they began to rotate faster and faster.

As each cloud fragment contracted, its temperature increased. Eventually, the temperature in the core of one of these cloud fragments reached about 10 million degrees Celsius and nuclear fusion began. A star was born—the beginning of the Sun.

It is unlikely that the Sun formed alone. A cluster of stars like the Sun, or smaller, likely formed from fragments of the original cloud. The Sun probably escaped from this cluster and has since revolved around the galaxy about 20 times.

Reading Check *What is nuclear fusion?*

Planet Formation Not all of the nearby gas, ice, and dust was drawn into the core of the cloud fragment. The matter that did not get pulled into the center collided and stuck together to form the planets and asteroids. Close to the Sun, the temperature was hot, and the easily vaporized elements could not condense into solids. This is why lighter elements are scarcer in the planets near the Sun than in planets farther out in the solar system.

The inner planets of the solar system—Mercury, Venus, Earth, and Mars—are small, rocky planets with iron cores. The outer planets are Jupiter, Saturn, Uranus, Neptune, and Pluto. Pluto, a small planet, is the only outer planet made mostly of rock and ice. The other outer planets are much larger and are made mostly of lighter substances such as hydrogen, helium, methane, and ammonia.

Figure 2 Systems of planets, such as the solar system, form in areas of space like this, called a nebula.

Figure 3

Through careful observations, astronomers have found clues that help explain how the solar system may have formed. **A** More than 4.6 billion years ago, the solar system was a cloud fragment of gas, ice, and dust. **B** Gradually, this cloud fragment contracted into a large, tightly packed, spinning disk. The disk's center was so hot and dense that nuclear fusion reactions began to occur, and the Sun was born. **C** Eventually, the rest of the material in the disk cooled enough to clump into scattered solids. **D** Finally, these clumps collided and combined to become the nine planets that make up the solar system today.

Table 1 Average Orbital Speed	
Planet	Average Orbital Speed (km/s)
Mercury	48
Venus	35
Earth	30
Mars	24
Jupiter	13
Saturn	9.7
Uranus	6.8
Neptune	5.4
Pluto	4.7

Johannes Kepler

Motions of the Planets

INTEGRATE Physics

When Nicholas Copernicus developed his Sun-centered model of the solar system, he thought that the planets orbited the Sun in circles. In the early 1600s, German mathematician Johannes Kepler began studying the orbits of the planets. He discovered that the shapes of the orbits are not circular. They are oval shaped, or elliptical. His calculations further showed that the Sun is not at the center of the orbits but is slightly offset.

Kepler also discovered that the planets travel at different speeds in their orbits around the Sun, as shown in **Table 1.** You can see that the planets closer to the Sun travel faster than planets farther away from the Sun. Because of their slower speeds and the longer distances they must travel, the outer planets take much longer to orbit the Sun than the inner planets do.

Copernicus's ideas, considered radical at the time, led to the birth of modern astronomy. Early scientists didn't have technology such as space probes to learn about the planets. Nevertheless, they developed theories about the solar system that still are used today.

section 1 review

Summary

Ideas About the Solar System
- The planets in the solar system revolve around the Sun.
- The Sun's immense gravity holds the planets in their orbits.

How the Solar System Formed
- The solar system formed from a piece of a nebula of gas, ice, and dust.
- As the piece of nebula contracted, nuclear fusion began at its center and the Sun was born.

Motion of the Planets
- The planets' orbits are elliptical.
- Planets that are closer to the Sun revolve faster than those that are farther away from the Sun.

Self Check

1. **Describe** the Sun-centered model of the solar system. What holds the solar system together?
2. **Explain** how the planets in the solar system formed.
3. **Infer** why life is unlikely on the outer planets in spite of the presence of water, methane, and ammonia—materials needed for life to develop.
4. **List** two reasons why the outer planets take longer to orbit the Sun than the inner planets do.
5. **Think Critically** Would a year on the planet Neptune be longer or shorter than an earth year? Explain.

Applying Skills

6. **Concept Map** Make a concept map that compares and contrasts the Earth-centered model with the Sun-centered model of the solar system.

Planetary Orbits

Planets travel around the Sun along paths called orbits. As you construct a model of a planetary orbit, you will observe that the shape of planetary orbits is an ellipse.

▶ Real-World Question

How can you model planetary orbits?

Goals
- ■ **Model** planetary orbits.
- ■ **Calculate** the eccentricity of ellipses.

Materials
thumbtacks or pins (2)   metric ruler
cardboard (23 cm × 30 cm)   string (25 cm)
paper (21.5 cm × 28 cm)   pencil

Safety Precautions

▶ Procedure

1. Place a blank sheet of paper on top of the cardboard and insert two thumbtacks or pins about 3 cm apart.

2. Tie the string into a circle with a circumference of 15 cm to 20 cm. Loop the string around the thumbtacks. With someone holding the tacks or pins, place your pencil inside the loop and pull it tight.

3. Moving the pencil around the tacks and keeping the string tight, mark a line until you have completed a smooth, closed curve.

4. Repeat steps 1 through 3 several times. First, vary the distance between the tacks, then vary the length of the string. However, change only one of these each time. Make a data table to record the changes in the sizes and shapes of the ellipses.

5. Orbits usually are described in terms of eccentricity, *e*, which is determined by dividing the distance, *d*, between the foci (fixed points—here, the tacks) by the length, *l*, of the major axis. See the diagram below.

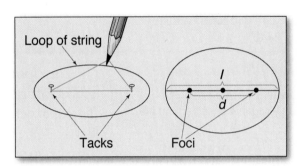

6. **Calculate** and record the eccentricity of the ellipses that you constructed.

7. **Research** the eccentricities of planetary orbits. Construct an ellipse with the same eccentricity as Earth's orbit.

▶ Conclude and Apply

1. **Analyze** the effect that a change in the length of the string or the distance between the tacks has on the shape of the ellipse.

2. **Hypothesize** what must be done to the string or placement of tacks to decrease the eccentricity of a constructed ellipse.

3. **Describe** the shape of Earth's orbit. Where is the Sun located within the orbit?

Compare your results with those of other students. **For more help, refer to the** Science Skill Handbook.

The Inner Planets

What You'll Learn

- **List** the inner planets in order from the Sun.
- **Describe** each inner planet.
- **Compare and contrast** Venus and Earth.

Why It's Important

The planet that you live on is uniquely capable of sustaining life.

🔎 Review Vocabulary

space probe: an instrument that is sent to space to gather information and send it back to Earth

New Vocabulary

- Mercury
- Earth
- Venus
- Mars

as you read

Inner Planets

Today, people know more about the solar system than ever before. Better telescopes allow astronomers to observe the planets from Earth and space. In addition, space probes have explored much of the solar system. Prepare to take a tour of the solar system through the eyes of some space probes.

Mercury

The closest planet to the Sun is **Mercury.** The first American spacecraft mission to Mercury was in 1974–1975 by *Mariner 10.* The spacecraft flew by the planet and sent pictures back to Earth. *Mariner 10* photographed only 45 percent of Mercury's surface, so scientists don't know what the other 55 percent looks like. What they do know is that the surface of Mercury has many craters and looks much like Earth's Moon. It also has cliffs as high as 3 km on its surface. These cliffs might have formed at a time when Mercury shrank in diameter, as seen in **Figure 4.**

Why would Mercury have shrunk? *Mariner 10* detected a weak magnetic field around Mercury. This indicates that the planet has an iron core. Some scientists hypothesize that Mercury's crust solidified while the iron core was still hot and molten. As the core started to solidify, it contracted. The cliffs resulted from breaks in the crust caused by this contraction.

Figure 4 Large cliffs on Mercury might have formed when the crust of the planet broke as the planet contracted.

Mercury has many craters.

Cliffs on the surface provide evidence that Mercury shrank.

Does Mercury have an atmosphere? Because of Mercury's low gravitational pull and high daytime temperatures, most gases that could form an atmosphere escape into space. *Mariner 10* found traces of hydrogen and helium gas that were first thought to be an atmosphere. However, these gases are now known to be temporarily taken from the solar wind.

Earth-based observations have found traces of sodium and potassium around Mercury. These atoms probably come from rocks in the planet's crust. Therefore, Mercury has no true atmosphere. This lack of atmosphere and its nearness to the Sun cause Mercury to have great extremes in temperature. Mercury's temperature can reach 425°C during the day, and it can drop to −170°C at night.

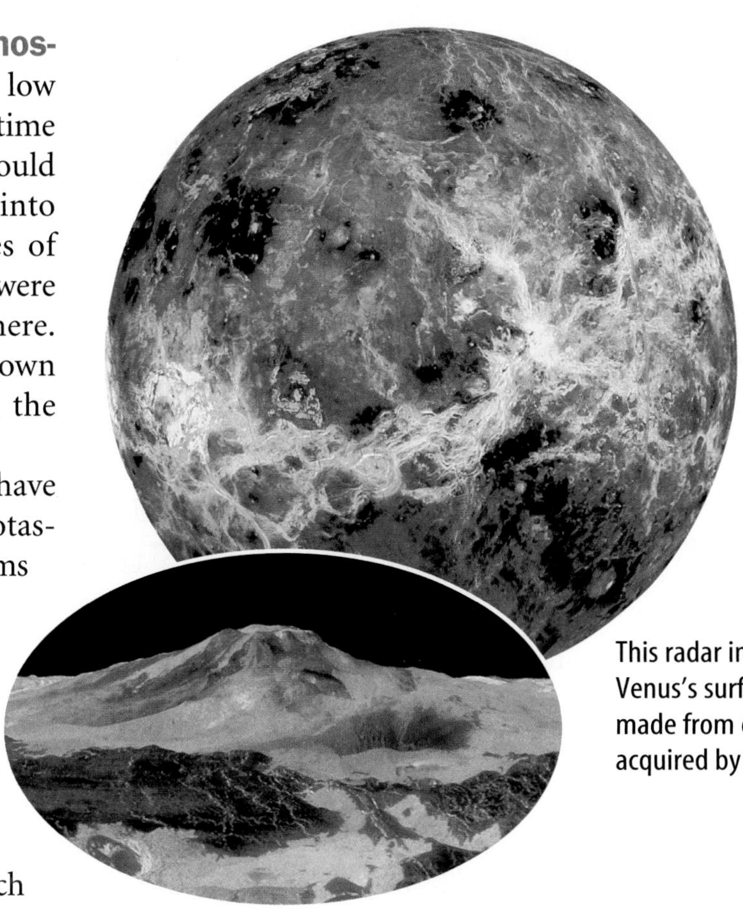

This radar image of Venus's surface was made from data acquired by *Magellan*.

Maat Mons is the highest volcano on Venus. Lava flows extend for hundreds of kilometers across the plains.

Venus

The second planet from the Sun is **Venus,** shown in **Figure 5.** Venus is sometimes called Earth's twin because its size and mass are similar to Earth's. In 1962, *Mariner 2* flew past Venus and sent back information about Venus's atmosphere and rotation. The former Soviet Union landed the first probe on the surface of Venus in 1970. *Venera 7*, however, stopped working in less than an hour because of the high temperature and pressure. Additional *Venera* probes photographed and mapped the surface of Venus. Between 1990 and 1994, the U.S. *Magellan* probe used its radar to make the most detailed maps yet of Venus's surface. It collected radar images of 98 percent of Venus's surface. Notice the huge volcano in **Figure 5.**

Clouds on Venus are so dense that only a small percentage of the sunlight that strikes the top of the clouds reaches the planet's surface. The sunlight that does get through warms Venus's surface, which then gives off heat to the atmosphere. Much of this heat is absorbed by carbon dioxide gas in Venus's atmosphere. This causes a greenhouse effect similar to, but more intense than, Earth's greenhouse effect. Due to this intense greenhouse effect, the temperature on the surface of Venus is between 450°C and 475°C.

Figure 5 Venus is the second planet from the Sun.

Earth

Figure 6 shows **Earth,** the third planet from the Sun. The average distance from Earth to the Sun is 150 million km, or one astronomical unit (AU). Unlike other planets, Earth has abundant liquid water and supports life. Earth's atmosphere causes most meteors to burn up before they reach the surface, and it protects life-forms from the effects of the Sun's intense radiation.

Figure 6 More than 70 percent of Earth's surface is covered by liquid water.
Explain *how Earth is unique.*

Mars

Look at **Figure 7.** Can you guess why **Mars,** the fourth planet from the Sun, is called the red planet? Iron oxide in soil on its surface gives it a reddish color. Other features visible from Earth are Mars's polar ice caps and changes in the coloring of the planet's surface. The ice caps are made of frozen water covered by a layer of frozen carbon dioxide.

Most of the information scientists have about Mars came from *Mariner 9,* the *Viking* probes, *Mars Pathfinder, Mars Global Surveyor,* and *Mars Odyssey. Mariner 9* orbited Mars in 1971 and 1972. It revealed long channels on the planet that might have been carved by flowing water. *Mariner 9* also discovered the largest volcano in the solar system, Olympus Mons, shown in **Figure 7.** Olympus Mons is probably extinct. Large rift valleys that formed in the Martian crust also were discovered. One such valley, Valles Marineris, is shown in **Figure 7.**

Figure 7 Many features on Mars are similar to those on Earth.

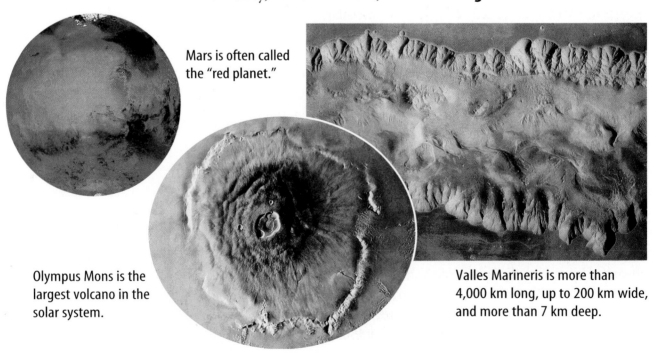

Mars is often called the "red planet."

Olympus Mons is the largest volcano in the solar system.

Valles Marineris is more than 4,000 km long, up to 200 km wide, and more than 7 km deep.

The *Viking* Probes The *Viking 1* and *2* probes arrived at Mars in 1976. Each spacecraft consisted of an orbiter and a lander. The *Viking 1* and *2* orbiters photographed the entire surface of Mars from their orbits, while the *Viking 1* and *2* landers touched down on the planet's surface. The landers carried equipment to detect possible life on Mars. Some of this equipment was designed to analyze gases given off by Martian soil. These experiments found no conclusive evidence of life on Mars.

Pathfinder, Global Surveyor, and Odyssey The *Mars Pathfinder* carried a robot rover named Sojourner with equipment that allowed it to analyze samples of Martian rock and soil. Data from these tests indicated that iron in Mars's crust might have been leached out by groundwater. Cameras onboard *Global Surveyor* showed features that look like gullies formed by flowing water and deposits of sediment carried by the water flows. The features, shown in **Figure 8,** are young enough that scientists are considering the idea that liquid groundwater might exist on Mars and that it sometimes reaches the surface. It is also possible that volcanic activity might melt frost beneath the Martian surface. The features compare to those formed by flash floods on Earth, such as on Mount St. Helens.

More recently, instruments on another probe called *Mars Odyssey* detected frozen water on Mars. The water occurs as frost beneath a thin layer of soil. The frost is common in the far northern and far southern parts of Mars.

✓ Reading Check *What evidence indicates that Mars has water?*

Inferring Effects of Gravity

Procedure
1. Suppose you are a crane operator who is sent to Mars to help build a Mars colony.
2. You know that your crane can lift 44,500 N on Earth and Mars, but the gravity on Mars is only 40 percent of Earth's gravity.
3. Determine how much mass your crane could lift on Earth and Mars.

Analysis
1. How can what you have discovered be an advantage over construction on Earth?
2. How might construction advantages change the overall design of the Mars colony?

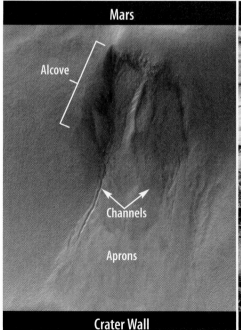

Mars

Alcove

Channels

Aprons

Crater Wall

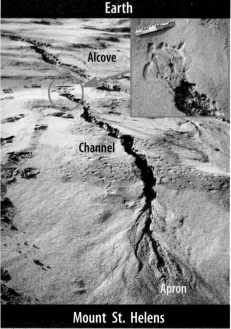

Earth

Alcove

Channel

Apron

Mount St. Helens

Figure 8 Compare the features found on Mars with those found on an area of Mount St. Helens in Washington state that experienced a flash flood.

Mars's Atmosphere The *Viking* and *Global Surveyor* probes analyzed gases in the Martian atmosphere and determined atmospheric pressure and temperature. They found that Mars's atmosphere is much thinner than Earth's. It is composed mostly of carbon dioxide, with some nitrogen and argon. Surface temperatures range from −125°C to 35°C. The temperature difference between day and night results in strong winds on the planet, which can cause global dust storms during certain seasons. This information will help in planning possible human exploration of Mars in the future.

Martian Seasons Mars's axis of rotation is tilted 25°, which is close to Earth's tilt of 23.5°. Because of this, Mars goes through seasons as it orbits the Sun, just like Earth does. The polar ice caps on Mars change with the season. During winter, carbon dioxide ice accumulates and makes the ice cap larger. During summer, carbon dioxide ice changes to carbon dioxide gas and the ice cap shrinks. As one ice cap gets larger, the other ice cap gets smaller. The color of the ice caps and other areas on Mars also changes with the season. The movement of dust and sand during dust storms causes the changing colors.

Applying Math   Use Percentages

DIAMETER OF MARS The diameter of Earth is 12,756 km. The diameter of Mars is 53.3 percent of the diameter of Earth. Calculate the diameter of Mars.

Solution

1 *This is what you know:*
- diameter of Earth: 12,756 km
- percent of Earth's diameter: 53.3%
- decimal equivalent: 0.533 (53.3% ÷ 100)

2 *This is what you need to find:*   diameter of Mars

3 *This is the procedure you need to use:*   Multiply the diameter of Earth by the decimal equivalent.
(12,756 km) × (0.533) = 6,799 km

Practice Problems

1. Use the same procedure to calculate the diameter of Venus. Its diameter is 94.9 percent of the diameter of Earth.

2. Calculate the diameter of Mercury. Its diameter is 38.2 percent of the diameter of Earth.

For more practice, visit bookj.msscience.com/math_practice

Martian Moons Mars has two small, irregularly shaped moons that are heavily cratered. Phobos, shown in **Figure 9,** is about 25 km in length, and Deimos is about 13 km in length. Deimos orbits Mars once every 31 h, while Phobos speeds around Mars once every 7 h.

Phobos has many interesting surface features. Grooves and chains of smaller craters seem to radiate out from the large Stickney Crater. Some of the grooves are 700 m across and 90 m deep. These features probably are the result of the large impact that formed the Stickney Crater.

Deimos is the outer of Mars's two moons. It is among the smallest known moons in the solar system. Its surface is smoother in appearance than that of Phobos because some of its craters have partially filled with soil and rock.

As you toured the inner planets through the eyes of the space probes, you saw how each planet is unique. Refer to **Table 3** following Section 3 for a summary of the planets. Mercury, Venus, Earth, and Mars are different from the outer planets, which you'll explore in the next section.

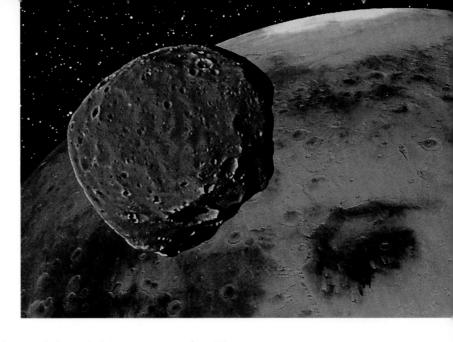

Figure 9 Phobos orbits Mars once every 7 h.
Infer *why Phobos has so many craters.*

section 2 review

Summary

Mercury
- Mercury is extremely hot during the day and extremely cold at night.
- Its surface has many craters.

Venus
- Venus's size and mass are similar to Earth's.
- Temperatures on Venus are between 450°C and 475°C.

Earth
- Earth is the only planet known to support life.

Mars
- Mars has polar ice caps, channels that might have been carved by water, and the largest volcano in the solar system, Olympus Mons.

Self Check

1. **Explain** why Mercury's surface temperature varies so much from day to night.
2. **List** important characteristics for each inner planet.
3. **Infer** why life is unlikely on Venus.
4. **Identify** the inner planet that is farthest from the Sun. Identify the one that is closest to the Sun.
5. **Think Critically** Aside from Earth, which inner planet could humans visit most easily? Explain.

Applying Math

6. **Use Statistics** The inner planets have the following average densities: Mercury, 5.43 g/cm³; Venus, 5.24 g/cm³; Earth, 5.52 g/cm³; and Mars, 3.94 g/cm³. Which planet has the highest density? Which has the lowest? Calculate the range of these data.

The Outer Planets

as you read

What **You'll Learn**

- **Describe** the characteristics of Jupiter, Saturn, Uranus, and Neptune.
- **Explain** how Pluto differs from the other outer planets.

Why **It's Important**

Studying the outer planets will help scientists understand Earth.

🔎 **Review Vocabulary**

moon: a natural satellite of a planet that is held in its orbit around the planet by the planet's gravitational pull

New Vocabulary

- Jupiter
- Great Red Spot
- Saturn
- Uranus
- Neptune
- Pluto

Outer Planets

You might have heard about *Voyager*, *Galileo*, and *Cassini*. They were not the first probes to the outer planets, but they gathered a lot of new information about them. Follow the space-crafts as you read about their journeys to the outer planets.

Jupiter

In 1979, *Voyager 1* and *Voyager 2* flew past **Jupiter,** the fifth planet from the Sun. *Galileo* reached Jupiter in 1995, and *Cassini* flew past Jupiter on its way to Saturn in 2000. The spacecrafts gathered new information about Jupiter. The *Voyager* probes revealed that Jupiter has faint dust rings around it and that one of its moons has active volcanoes on it.

Jupiter's Atmosphere Jupiter is composed mostly of hydrogen and helium, with some ammonia, methane, and water vapor. Scientists hypothesize that the atmosphere of hydrogen and helium changes to an ocean of liquid hydrogen and helium toward the middle of the planet. Below this liquid layer might be a rocky core. The extreme pressure and temperature, however, would make the core different from any rock on Earth.

You've probably seen pictures from the probes of Jupiter's colorful clouds. In **Figure 10,** you can see bands of white, red, tan, and brown clouds in its atmosphere. Continuous storms of swirling, high-pressure gas have been observed on Jupiter. The **Great Red Spot** is the most spectacular of these storms.

Figure 10 Jupiter is the largest planet in the solar system.

Notice the colorful bands of clouds in Jupiter's atmosphere.

The Great Red Spot is a giant storm about 25,000 km in size from east to west.

Table 2 Large Moons of Jupiter

Io The most volcanically active object in the solar system; sulfurous compounds give it its distinctive reddish and orange colors; has a thin oxygen, sulfur, and sulfur dioxide atmosphere.

Europa Rocky interior is covered by a 100-km-thick crust of ice, which has a network of cracks; an ocean might exist under the ice crust; has a thin oxygen atmosphere.

Ganymede Has a crust of ice about 500 km thick, covered with grooves; crust might surround an ocean of water or slushy ice; has a rocky core and a thin oxygen atmosphere.

Callisto Has a heavily cratered crust of ice and rock several hundred kilometers thick; crust might surround a salty ocean around a rock core; has a thin atmosphere of carbon dioxide.

Moons of Jupiter At least 61 moons orbit Jupiter. In 1610, the astronomer Galileo Galilei was the first person to see Jupiter's four largest moons, shown in **Table 2.** Io (I oh) is the closest large moon to Jupiter. Jupiter's tremendous gravitational force and the gravity of Europa, Jupiter's next large moon, pull on Io. This force heats up Io, causing it to be the most volcanically active object in the solar system. You can see a volcano erupting on Io in **Figure 11.** Europa is composed mostly of rock with a thick, smooth crust of ice. Under the ice might be an ocean as deep as 200 km. If this ocean of water exists, it will be the only place in the solar system, other than Earth and possibly Ganymede and Callisto, where liquid water exists in large quantities. Next is Ganymede, the largest moon in the solar system—larger even than the planet Mercury. Callisto, the last of Jupiter's large moons, is composed mostly of ice and rock. Studying these moons adds to knowledge about the origin of Earth and the rest of the solar system.

Figure 11 *Voyager 2* photographed the eruption of this volcano on Io in July 1979.

Figure 12 Saturn's rings are composed of pieces of rock and ice.

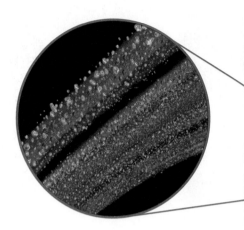

Saturn

The *Voyager* probes next surveyed Saturn in 1980 and 1981. *Cassini* is scheduled to reach Saturn on July 1, 2004. **Saturn** is the sixth planet from the Sun. It is the second-largest planet in the solar system, but it has the lowest density.

Saturn's Atmosphere Similar to Jupiter, Saturn is a large, gaseous planet. It has a thick outer atmosphere composed mostly of hydrogen and helium. Saturn's atmosphere also contains ammonia, methane, and water vapor. As you go deeper into Saturn's atmosphere, the gases gradually change to liquid hydrogen and helium. Below its atmosphere and liquid layer, Saturn might have a small, rocky core.

Rings and Moons The *Voyager* probes gathered new information about Saturn's ring system. The probes showed that Saturn has several broad rings. Each large ring is composed of thousands of thin ringlets. **Figure 12** shows that Saturn's rings are composed of countless ice and rock particles. These particles range in size from a speck of dust to tens of meters across. Saturn's ring system is the most complex one in the solar system.

At least 31 moons orbit Saturn. Saturn's gravity holds these moons in their orbits around Saturn, just like the Sun's gravity holds the planets in their orbits around the Sun. The largest of Saturn's moons, Titan, is larger than the planet Mercury. It has an atmosphere of nitrogen, argon, and methane. Thick clouds make it difficult for scientists to study the surface of Titan.

Uranus

Beyond Saturn, *Voyager 2* flew by Uranus in 1986. **Uranus** (YOOR uh nus) is the seventh planet from the Sun and was discovered in 1781. It is a large, gaseous planet with at least 21 moons and a system of thin, dark rings. Uranus's largest moon, Titania, has many craters and deep valleys. The valleys on this moon indicate that some process reshaped its surface after it formed. Uranus's 11 rings surround the planet's equator.

Uranus's Characteristics The atmosphere of Uranus is composed of hydrogen, helium, and some methane. Methane gives the planet the bluish-green color that you see in **Figure 13.** Methane absorbs the red and yellow light, and the clouds reflect the green and blue. Few cloud bands and storm systems can be seen on Uranus. Evidence suggests that under its atmosphere, Uranus has a mantle of liquid water, methane, and ammonia surrounding a rocky core.

Figure 14 shows one of the most unusual features of Uranus. Its axis of rotation is tilted on its side compared with the other planets. The axes of rotation of the other planets, except Pluto, are nearly perpendicular to the planes of their orbits. However, Uranus's axis of rotation is nearly parallel to the plane of its orbit. Some scientists believe a collision with another object tipped Uranus on its side.

Figure 13 The atmosphere of Uranus gives the planet its distinct bluish-green color.

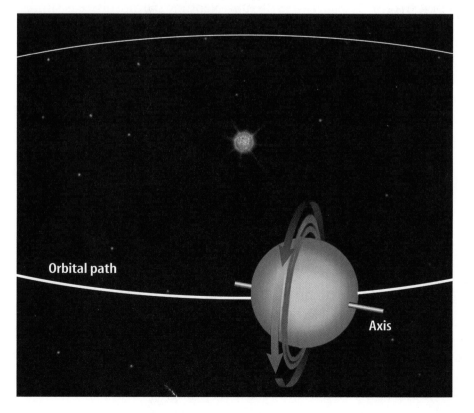

Orbital path

Axis

Figure 14 Uranus's axis of rotation is nearly parallel to the plane of its orbit. During its revolution around the Sun, each pole, at different times, points almost directly at the Sun.

Neptune has a distinctive bluish-green color.

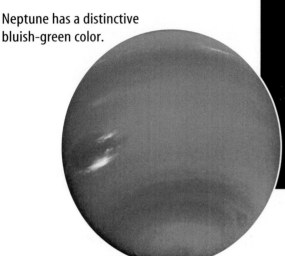

The pinkish hue of Neptune's largest moon, Triton, is thought to come from an evaporating layer of nitrogen and methane ice.

Figure 15 Neptune is the eighth planet from the Sun.

Names of Planets The names of most of the planets in the solar system come from Roman or Greek mythology. For example, Neptune was the Roman god of the sea, and Pluto was the Greek god of the underworld. Research to learn about the names of the other planets. Write a paragraph in your Science Journal that summarizes what you learn.

Neptune

Passing Uranus, *Voyager 2* traveled to Neptune, another large, gaseous planet. Discovered in 1846, **Neptune** is usually the eighth planet from the Sun. However, Pluto's orbit crosses inside Neptune's during part of its voyage around the Sun. Between 1979 and 1999, Pluto was closer to the Sun than was Neptune.

Neptune's Characteristics Neptune's atmosphere is similar to Uranus's atmosphere. The methane content gives Neptune, shown in **Figure 15,** its distinctive bluish-green color, just as it does for Uranus.

☑ **Reading Check** *What gives Neptune its bluish-green color?*

Neptune has dark-colored storms in its atmosphere that are similar to the Great Red Spot on Jupiter. One discovered by *Voyager 2* in 1989 was called the Great Dark Spot. It was about the size of Earth. However, observations by the *Hubble Space Telescope* in 1994 showed that the Great Dark Spot had disappeared. Bright clouds also form and then disappear. This shows that Neptune's atmosphere is active and changes rapidly.

Under its atmosphere, Neptune is thought to have a layer of liquid water, methane, and ammonia that might change to solid ice. Neptune probably has a rocky core.

Neptune has at least 11 moons and several rings. Triton, shown in **Figure 15,** is Neptune's largest moon. It has a thin atmosphere composed mostly of nitrogen. Neptune's rings are young and probably won't last very long.

Pluto

The smallest planet in the solar system, and the one scientists know the least about, is **Pluto.** During part of its 248-year orbit, Pluto is closer to the Sun than Neptune. However, because Pluto is farther from the Sun than Neptune during most of its orbit, it is considered to be the ninth planet from the Sun. Pluto is vastly different from the other outer planets. It's surrounded by only a thin atmosphere, and it's the only outer planet with a solid, icy-rock surface.

Pluto's Moon Pluto's single moon, Charon, has a diameter about half the size of Pluto's. It was discovered in 1978 when a bulge was noticed on a photograph of the planet. Later photographs, taken with improved telescopes, showed that the bulge was a moon. Pluto and Charon are shown in **Figure 16.** Because of their close size and orbit, some scientists consider them to be a double planet.

Data from the *Hubble Space Telescope* indicate the presence of a vast disk of icy objects called the Kuiper Belt beyond Neptune's orbit. Some of the objects are hundreds of kilometers in diameter. One found in 2002 is about 1,300 km in diameter. Many astronomers think that Pluto and Charon are members of the Kuiper Belt.

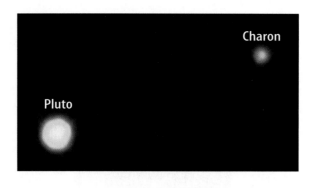

Figure 16 The *Hubble Space Telescope* gave astronomers their first clear view of Pluto and Charon as distinct objects.

section 3 review

Summary

Jupiter
- Jupiter is the largest planet in the solar system.
- The Great Red Spot is a huge storm on Jupiter.

Saturn
- Saturn has a complex system of rings.

Uranus
- Uranus has a bluish-green color caused by methane in its atmosphere.

Neptune
- Like Uranus, Neptune has a bluish-green color.
- Neptune's atmosphere has storms and can change rapidly.

Pluto
- Pluto is a small, ice-rock planet.
- Its moon, Charon, is about half as large as Pluto.

Self Check

1. **Describe** the differences between the outer planets and the inner planets.
2. **Describe** what Saturn's rings are made from.
3. **Explain** how Pluto is a unique outer planet.
4. **Explain** how Uranus's axis of rotation differs from those of most other planets.
5. **Think Critically** What would seasons be like on Uranus? Explain.

Applying Skills

6. **Identify a Question** Pluto has never been visited by a space probe, so many questions are unanswered. Think of a question about Pluto that you'd like to have answered. Then, explain how the answer to your question could be discovered.

Table 3 Planets

Mercury

- closest to the Sun
- second-smallest planet
- surface has many craters and high cliffs
- no atmosphere
- temperatures range from 425°C during the day to −170°C at night
- has no moons

Venus

- similar to Earth in size and mass
- thick atmosphere made mostly of carbon dioxide
- droplets of sulfuric acid in atmosphere give clouds a yellowish color
- surface has craters, faultlike cracks, and volcanoes
- greenhouse effect causes surface temperatures of 450°C to 475°C
- has no moons

Earth

- atmosphere protects life
- surface temperatures allow water to exist as solid, liquid, and gas
- only planet where life is known to exist
- has one large moon

Mars

- surface appears reddish-yellow because of iron oxide in soil
- ice caps are made of frozen carbon dioxide and water
- channels indicate that water had flowed on the surface; has large volcanoes and valleys
- has a thin atmosphere composed mostly of carbon dioxide
- surface temperatures range from −125°C to 35°C
- huge dust storms often blanket the planet
- has two small moons

Table 3 Planets

Jupiter
- largest planet
- has faint rings
- atmosphere is mostly hydrogen and helium; continuous storms swirl on the planet—the largest is the Great Red Spot
- has four large moons and at least 57 smaller moons; one of its moons, Io, has active volcanoes

Saturn
- second-largest planet
- thick atmosphere is mostly hydrogen and helium
- has a complex ring system
- has at least 31 moons—the largest, Titan, is larger than Mercury

Uranus
- large, gaseous planet with thin, dark rings
- atmosphere is hydrogen, helium, and methane
- axis of rotation is nearly parallel to plane of orbit
- has at least 21 moons

Neptune
- large, gaseous planet with rings that vary in thickness
- is sometimes farther from the Sun than Pluto is
- methane atmosphere causes its bluish-green color
- has dark-colored storms in atmosphere
- has at least 11 moons

Pluto
- small, icy-rock planet with thin atmosphere
- single moon, Charon, is half the diameter of the planet

Other Objects in the Solar System

as you read

What You'll Learn

- **Describe** how comets change when they approach the Sun.
- **Distinguish** among comets, meteoroids, and asteroids.
- **Explain** that objects from space sometimes impact Earth.

Why It's Important

Comets, asteroids, and most meteorites are very old. Scientists can learn about the early solar system by studying them.

Review Vocabulary

crater: a nearly circular depression in a planet, moon, or asteroid that formed when an object from space hit its surface

New Vocabulary

- comet
- meteorite
- meteor
- asteroid

Comets

The planets and their moons are the most noticeable members of the Sun's family, but many other objects also orbit the Sun. Comets, meteoroids, and asteroids are other important objects in the solar system.

You might have heard of Halley's comet. A **comet** is composed of dust and rock particles mixed with frozen water, methane, and ammonia. Halley's comet was last seen from Earth in 1986. English astronomer Edmund Halley realized that comet sightings that had taken place about every 76 years were really sightings of the same comet. This comet, which takes about 76 years to orbit the Sun, was named after him.

Oort Cloud Astronomer Jan Oort proposed the idea that billions of comets surround the solar system. This cloud of comets, called the Oort Cloud, is located beyond the orbit of Pluto. Oort suggested that the gravities of the Sun and nearby stars interact with comets in the Oort Cloud. Comets either escape from the solar system or get captured into smaller orbits.

Comet Hale-Bopp On July 23, 1995, two amateur astronomers made an exciting discovery. A new comet, Comet Hale-Bopp, was headed toward the Sun. Larger than most that approach the Sun, it was the brightest comet visible from Earth in 20 years. Shown in **Figure 17,** Comet Hale-Bopp was at its brightest in March and April 1997.

Figure 17 Comet Hale-Bopp was most visible in March and April 1997.

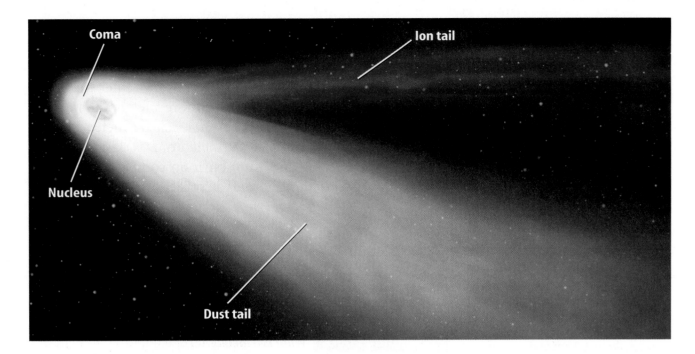

Coma

Ion tail

Nucleus

Dust tail

Structure of Comets

The *Hubble Space Telescope* and space-crafts such as the *International Cometary Explorer* have gathered information about comets. In 2006, a spacecraft called *Stardust* will return a capsule to Earth containing samples of dust from a comet's tail. Notice the structure of a comet shown in **Figure 18.** It is a mass of frozen ice and rock similar to a large, dirty snowball.

As a comet approaches the Sun, it changes. Ices of water, methane, and ammonia vaporize because of the heat from the Sun, releasing dust and bits of rock. The gases and released dust form a bright cloud called a coma around the nucleus, or solid part, of the comet. The solar wind pushes on the gases and dust in the coma, causing the particles to form separate tails that point away from the Sun.

After many trips around the Sun, most of the ice in a comet's nucleus has vaporized. All that's left are dust and rock, which are spread throughout the orbit of the original comet.

Meteoroids, Meteors, and Meteorites

You learned that comets vaporize and break up after they have passed close to the Sun many times. The small pieces from the comet's nucleus spread out into a loose group within the original orbit of the comet. These pieces of dust and rock, along with those derived from other sources, are called meteoroids.

Sometimes the path of a meteoroid crosses the position of Earth, and it enters Earth's atmosphere at speeds of 15 km/s to 70 km/s. Most meteoroids are so small that they completely burn up in Earth's atmosphere. A meteoroid that burns up in Earth's atmosphere is called a **meteor,** shown in **Figure 19.**

Figure 18 A comet consists of a nucleus, a coma, a dust tail, and an ion tail.

Figure 19 A meteoroid that burns up in Earth's atmosphere is called a meteor.

Figure 20 Meteorites occasionally strike Earth's surface. A large meteorite struck Arizona, forming a crater about 1.2 km in diameter and about 200 m deep.

Meteor Showers Each time Earth passes through the loose group of particles within the old orbit of a comet, many small particles of rock and dust enter the atmosphere. Because more meteors than usual are seen, the event is called a meteor shower.

When a meteoroid is large enough, it might not burn up completely in the atmosphere. If it strikes Earth, it is called a **meteorite.** Barringer Crater in Arizona, shown in **Figure 20,** was formed when a large meteorite struck Earth about 50,000 years ago. Most meteorites are probably debris from asteroid collisions or broken-up comets, but some originate from the Moon and Mars.

Reading Check *What is a meteorite?*

Asteroids

An **asteroid** is a piece of rock similar to the material that formed into the planets. Most asteroids are located in an area between the orbits of Mars and Jupiter called the asteroid belt. Find the asteroid belt in **Figure 21.** Why are they located there? The gravity of Jupiter might have kept a planet from forming in the area where the asteroid belt is located now.

Other asteroids are scattered throughout the solar system. They might have been thrown out of the belt by Jupiter's gravity. Some of these asteroids have orbits that cross Earth's orbit. Scientists monitor the positions of these asteroids. However, it is unlikely that an asteroid will hit Earth in the near future.

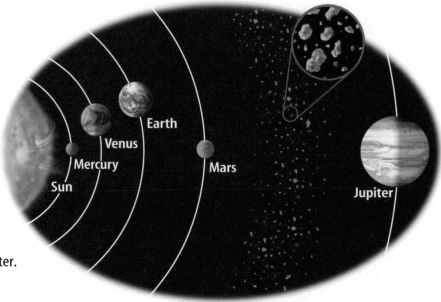

Figure 21 The asteroid belt lies between the orbits of Mars and Jupiter.

Exploring Asteroids The sizes of the asteroids in the asteroid belt range from tiny particles to objects 940 km in diameter. Ceres is the largest and the first one discovered. The next three in order of size are Vesta (530 km), Pallas (522 km), and 10 Hygiea (430 km). The asteroid Gaspra, shown in **Figure 22,** was photographed by *Galileo* on its way to Jupiter.

NEAR On February 14, 2000, the *Near Earth Asteroid Rendezvous (NEAR)* spacecraft went into orbit around the asteroid 433 Eros and successfully began its one-year mission of data gathering. Data from the spacecraft show that Eros's surface has a large number of craters. Other data indicate that Eros might be similar to the most common type of meteorite that strikes Earth. On February 12, 2001, *NEAR* ended its mission by becoming the first spacecraft to land softly on an asteroid.

Comets, asteroids, and most meteorites formed early in the history of the solar system. Scientists study these space objects to learn what the solar system might have been like long ago. Understanding this could help scientists better understand how Earth formed.

Figure 22 The asteroid Gaspra is about 20 km long.

---

section 4 review

Summary

Comets
- Comets consist of dust, rock, and different types of ice.
- Billions of comets surround the solar system in the Oort Cloud.

Meteoroids, Meteors, Meteorites
- When meteoroids burn up in the atmosphere, they are called meteors.
- Meteor showers occur when Earth crosses the orbital path of a comet.

Asteroids
- Many asteroids occur between the orbits of Mars and Jupiter. This region is called the asteroid belt.

Self Check

1. **Describe** how a comet changes when it comes close to the Sun.
2. **Explain** how craters form.
3. **Summarize** the differences between comets and asteroids.
4. **Describe** the mission of the *NEAR* spacecraft.
5. **Think Critically** A meteorite found in Antarctica is thought to have come from Mars. How could a rock from Mars get to Earth?

Applying Math

6. **Use Proportions** During the 2001 Leonid Meteor Shower, some people saw 20 meteors each minute. Assuming a constant rate, how many meteors did these people see in one hour?

Model and Invent

S☀lar System Distance Model

● Real-World Question

Distances between the Sun and the planets of the solar system are large. These large distances can be difficult to visualize. Can you design and create a model that will demonstrate the distances in the solar system?

Goals

■ **Design** a table of scale distances and model the distances between and among the Sun and the planets.

Possible Materials

meterstick
scissors
pencil
string (several meters)
notebook paper (several sheets)

Safety Precautions

Use care when handling scissors.

● Make a Model

1. **List** the steps that you need to take to make your model. Describe exactly what you will do at each step.

2. **List** the materials that you will need to complete your model.

3. **Describe** the calculations that you will use to get scale distances from the Sun for all nine planets.

4. **Make** a table of scale distances that you will use in your model. Show your calculations in your table.

5. **Write** a description of how you will build your model. Explain how it will demonstrate relative distances between and among the Sun and planets of the solar system.

Planetary Distances				
Planet	Distance to Sun (km)	Distance to Sun (AU)	Scale Distance (1 AU = 10 cm)	Scale Distance (1 AU = 2 cm)
Mercury	5.97×10^7	0.39		
Venus	1.08×10^8	0.72		
Earth	1.50×10^8	1.00		
Mars	2.28×10^8	1.52		Do not write in this book.
Jupiter	7.78×10^8	5.20		
Saturn	1.43×10^9	9.54		
Uranus	2.87×10^9	19.19		
Neptune	4.50×10^9	30.07		
Pluto	5.92×10^9	39.48		

▶ Test Your Model

1. **Compare** your scale distances with those of other students. Discuss why each of you chose the scale that you did.

2. Make sure your teacher approves your plan before you start.

3. **Construct** the model using your scale distances.

4. While constructing the model, write any observations that you or other members of your group make, and complete the data table in your Science Journal. Calculate the scale distances that would be used in your model if 1 AU = 2 m.

▶ Analyze Your Data

1. **Explain** how a scale distance is determined.

2. Was it possible to work with your scale? Explain why or why not.

3. How much string would be required to construct a model with a scale distance of 1 AU = 2 m?

4. Proxima Centauri, the closest star to the Sun, is about 270,000 AU from the Sun. Based on your scale, how much string would you need to place this star on your model?

▶ Conclude and Apply

1. **Summarize** your observations about distances in the solar system. How are distances between the inner planets different from distances between the outer planets?

2. Using your scale distances, determine which planet orbits closest to Earth. Which planet's orbit is second closest?

𝒞ommunicating Your Data

Compare your scale model with those of other students. Discuss any differences. **For more help, refer to the** Science Skill Handbook.

IT CAME FROM OUTER SPACE!

On September 4, 1990, Frances Pegg was unloading bags of groceries in her kitchen in Burnwell, Kentucky. Suddenly, she heard a loud crashing sound. Her husband Arthur heard the same sound. The sound frightened the couple's goat and horse. The noise had come from an object that had crashed through the Pegg's roof, their ceiling, and the floor of their porch. They couldn't see what the object was, but the noise sounded like a gunshot, and pieces of wood from their home flew everywhere. The next day the couple looked under their front porch and found the culprit—a chunk of rock from outer space. It was a meteorite.

For seven years, the Peggs kept their "space rock" at home, making them local celebrities. The rock appeared on TV, and the couple was interviewed by newspaper reporters. In 1997, the Peggs sold the meteorite to the National Museum of Natural History in Washington, D.C., which has a collection of more than 9,000 meteorites. Scientists there study meteorites to learn more about the solar system. One astronomer explained, "Meteorites formed at about the same time as the solar system, about 4.6 billion years ago, though some are younger."

Scientists especially are interested in the Burnwell meteorite because its chemical make up is different from other meteorites previously studied. The Burnwell meteorite is richer in metallic iron and nickel than other known meteorites and is less rich in some metals such as cobalt. Scientists are comparing the rare Burnwell rock with other data to find out if there are more meteorites like the one that fell on the Peggs' roof. But so far, it seems the Peggs' visitor from outer space is one-of-a-kind.

The Burnwell meteorite crashed into the Peggs' home and landed in their basement on the right.

Research Do research to learn more about meteorites. How do they give clues to how our solar system formed? Report to the class.

Science online
For more information, visit bookj.msscience.com/oops

Reviewing Main Ideas

Section 1 The Solar System

1. Early Greek scientists thought that Earth was at the center of the solar system. They thought that the planets and stars circled Earth.

2. Today, people know that objects in the solar system revolve around the Sun.

Section 2 The Inner Planets

1. The inner planets are Mercury, Venus, Earth, and Mars.

2. The inner planets are small, rocky planets.

Section 3 The Outer Planets

1. The outer planets are Jupiter, Saturn, Uranus, Neptune, and Pluto.

2. Pluto is a small icy planet. The other outer planets are large, gaseous planets.

Section 4 Other Objects in the Solar System

1. Comets are masses of ice and rock. When a comet approaches the Sun, some ice turns to gas and the comet glows brightly.

2. Meteors occur when small pieces of rock enter Earth's atmosphere and burn up.

Visualizing Main Ideas

Copy and complete the following concept map about the solar system.

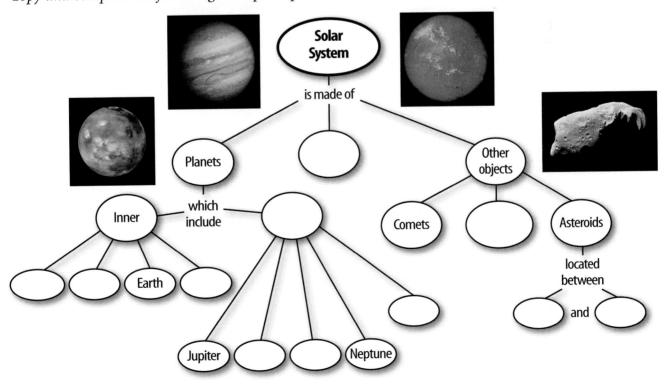

Using Vocabulary

asteroid p. 92
comet p. 90
Earth p. 78
Great Red Spot p. 82
Jupiter p. 82
Mars p. 78
Mercury p. 76
meteor p. 91
meteorite p. 92
Neptune p. 86
Pluto p. 87
Saturn p. 84
solar system p. 71
Uranus p. 85
Venus p. 77

Fill in the blanks with the correct words.

1. A meteoroid that burns up in Earth's atmosphere is called a(n) _____.

2. The Great Red Spot is a giant storm on _____.

3. _____ is the second largest planet.

4. The *Viking* landers tested for life on _____.

5. The _____ includes the Sun, planets, moons, and other objects.

Checking Concepts

Choose the word or phrase that best answers the question.

6. Who proposed a Sun-centered solar system?
   A) Ptolemy
   B) Copernicus
   C) Galileo
   D) Oort

7. What is the shape of planetary orbits?
   A) circles
   B) ellipses
   C) squares
   D) rectangles

8. Which planet has extreme temperatures because it has no atmosphere?
   A) Earth
   B) Jupiter
   C) Saturn
   D) Mercury

9. Where is the largest volcano in the solar system?
   A) Earth
   B) Jupiter
   C) Mars
   D) Uranus

Use the photo below to answer question 10.

10. Which planet has a complex ring system consisting of thousands of ringlets?
    A) Pluto
    B) Saturn
    C) Uranus
    D) Mars

11. What is a rock from space that strikes Earth's surface?
    A) asteroid
    B) meteoroid
    C) meteorite
    D) meteor

12. By what process does the Sun produce energy?
    A) magnetism
    B) nuclear fission
    C) nuclear fusion
    D) gravity

13. In what direction do comet tails point?
    A) toward the Sun
    B) away from the Sun
    C) toward Earth
    D) away from the Oort Cloud

14. Which planet has abundant surface water and is known to have life?
    A) Mars
    B) Jupiter
    C) Earth
    D) Pluto

15. Which planet has the highest temperatures because of the greenhouse effect?
    A) Mercury
    B) Venus
    C) Saturn
    D) Earth

Science Online bookj.msscience.com/vocabulary_puzzlemaker

Thinking Critically

16. Infer Why are probe landings on Jupiter not possible?

17. Concept Map Copy and complete the concept map on this page to show how a comet changes as it travels through space.

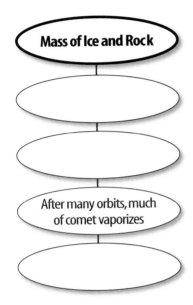

Mass of Ice and Rock

After many orbits, much of comet vaporizes

18. Recognize Cause and Effect What evidence suggests that liquid water is or once was present on Mars?

19. Venn Diagram Create a Venn diagram for Earth and Venus. Create a second Venn diagram for Uranus and Neptune. Which two planets do you think are more similar?

20. Recognize Cause and Effect Mercury is closer to the Sun than Venus, yet Venus has higher temperatures. Explain.

21. Make Models Make a model that includes the Sun, Earth, and the Moon. Use your model to demonstrate how the Moon revolves around Earth and how Earth and the Moon revolve around the Sun.

22. Form Hypotheses Why do Mars's two moons look like asteroids?

Performance Activities

23. Display Mercury, Venus, Mars, Jupiter, and Saturn can be observed with the unaided eye. Research when and where in the sky these planets can be observed during the next year. Make a display illustrating your findings. Take some time to observe some of these planets.

24. Short Story Select one of the planets or a moon in the solar system. Write a short story from the planet's or moon's perspective. Include scientifically correct facts and concepts in your story.

Applying Math

25. Saturn's Atmosphere Saturn's atmosphere consists of 96.3% hydrogen and 3.25% helium. What percentage of Saturn's atmosphere is made up of other gases?

26. Length of Day on Pluto A day on Pluto lasts 6.39 times longer than a day on Earth. If an Earth day lasts 24 h, how many hours is a day on Pluto?

Use the graph below to answer question 27.

Weight on Several Planets		
Planet	**Proportion of Earth's Gravity**	**Melissa's Weight (lbs)**
Mercury	0.378	
Venus	0.903	
Earth	1.000	70
Mars	0.379	
Jupiter	2.54	
Pluto	0.061	

27. Gravity and Weight Melissa weighs 70 lbs on Earth. Multiply Melissa's weight by the proportion of Earth's gravity for each planet to find out how much Melissa would weigh on each.

Part 1 Multiple Choice

Record your answers on the answer sheet provided by your teacher or on a sheet of paper.

Use the photo below to answer question 1.

1. What is shown in the photo above?
   A. asteroids   C. meteors
   B. comets   D. meteorites

2. Which is the ninth planet from the Sun?
   A. Earth   C. Jupiter
   B. Mars   D. Pluto

3. What is the name of Pluto's moon?
   A. Ganymede   C. Charon
   B. Titan   D. Phobos

4. Which object's gravity holds the planets in their orbits?
   A. Gaspra   C. Mercury
   B. Earth   D. the Sun

5. Which of the following occurs in a cycle?
   A. sunspot maxima and minima
   B. condensation of a nebula
   C. formation of a crater
   D. formation of a black hole

Test-Taking Tip

No Peeking During the test, keep your eyes on your own paper. If you need to rest them, close them or look up at the ceiling.

6. Which planet likely will be visited by humans in the future?
   A. Jupiter   C. Mars
   B. Venus   D. Neptune

7. Between which two planets' orbits does the asteroid belt occur?
   A. Mercury and Venus
   B. Earth and Mars
   C. Uranus and Neptune
   D. Mars and Jupiter

8. Who discovered that planets have elliptical orbits?
   A. Galileo Galilei
   B. Johannes Kepler
   C. Albert Einstein
   D. Nicholas Copernicus

Use the illustration below to answer question 9.

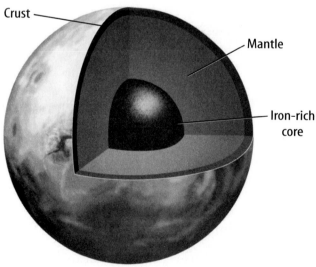

Crust — Mantle — Iron-rich core

0 1,000 km

9. Which of the following answers is a good estimate for the diameter of Mars?
   A. 23,122 km   C. 1,348 km
   B. 6,794 km   D. 12,583 km

Part 2 | Short Response/Grid In

Record your answers on the answer sheet provided by your teacher or on a sheet of paper.

10. Why does a moon remain in orbit around a planet?

11. Compare and contrast the inner planets and the outer planets.

12. Describe Pluto's surface. How is it different from the other outer planets?

13. Describe Saturn's rings. What are they made of?

14. What is the Great Red Spot?

15. How is Earth different from the other planets in the solar system?

Use the graph below to answwfer questions 16–19.

Viking Lander 1 Temperature Data

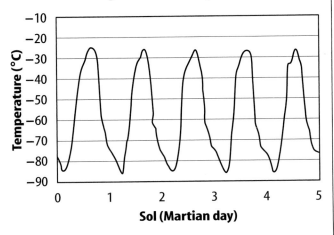

16. Why do the temperatures in the graph vary in a pattern?

17. Approximate the typical high temperature value measured by *Viking I*.

18. Approximate the typical low temperature value measured by *Viking I*.

19. What is the range of these temperature values?

Part 3 | Open Ended

Record your answers on a sheet of paper.

20. How might near-Earth-asteroids affect life on Earth? Why do astronomers search for them and monitor their positions?

Use the illustration below to answer question 21.

21. Explain how scientists hypothesize that the large cliffs on Mercury formed.

22. Describe the Sun-centered model of the solar system. How is it different from the Earth-centered model?

23. What is an astronomical unit? Why is it useful?

24. Compare and contrast the distances between the planets in the solar system. Which planets are relatively close together? Which planets are relatively far apart?

25. Summarize the current hypothesis about how the solar system formed.

26. Explain how Earth's gravity affects objects that are on or near Earth.

27. Describe the shape of planets' orbits. What is the name of this shape? Where is the Sun located?

28. Describe Jupiter's atmosphere. What characteristics can be observed in images acquired by space probes?

Stars and Galaxies

What's your address?

You know your address at home. You also know your address at school. But do you know your address in space? You live on a planet called Earth that revolves around a star called the Sun. Earth and the Sun are part of a galaxy called the Milky Way. It looks similar to galaxy M83, shown in the photo.

Science Journal Write a description in your Science Journal of the galaxy shown on this page.

Start-Up Activities

Why do clusters of galaxies move apart?

Astronomers know that most galaxies occur in groups of galaxies called clusters. These clusters are moving away from each other in space. The fabric of space is stretching like an inflating balloon. 👓

1. Partially inflate a balloon. Use a piece of string to seal the neck.

2. Draw six evenly spaced dots on the balloon with a felt-tipped marker. Label the dots A through F.

3. Use string and a ruler to measure the distance, in millimeters, from dot A to each of the other dots.

4. Inflate the balloon more.

5. Measure the distances from dot A again.

6. Inflate the balloon again and make new measurements.

7. **Think Critically** Imagine that each dot represents a cluster of galaxies and that the balloon represents the universe. Describe the motion of the clusters in your Science Journal.

 **Preview this chapter's content and activities at** bookj.msscience.com

Stars, Galaxies, and the Universe Make the following Foldable to show what you know about stars, galaxies, and the universe.

STEP 1 Fold a sheet of paper from side to side. Make the front edge about 1.25 cm shorter than the back edge.

STEP 2 Turn lengthwise and fold into thirds.

STEP 3 Unfold and cut only the top layer along both folds to make three tabs.

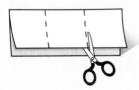

STEP 4 Label the tabs *Stars, Galaxies,* and *Universe.*

Read and Write Before you read the chapter, write what you already know about stars, galaxies, and the universe. As you read the chapter, add to or correct what you have written under the tabs.

as you read

What You'll Learn

- **Explain** why some constellations are visible only during certain seasons.
- **Distinguish** between absolute magnitude and apparent magnitude.

Why It's Important

The Sun is a typical star.

Review Vocabulary

star: a large, spherical mass of gas that gives off light and other types of radiation

New Vocabulary

- constellation
- absolute magnitude
- apparent magnitude
- light-year

Constellations

It's fun to look at clouds and find ones that remind you of animals, people, or objects that you recognize. It takes more imagination to play this game with stars. Ancient Greeks, Romans, and other early cultures observed patterns of stars in the night sky called **constellations.** They imagined that the constellations represented mythological characters, animals, or familiar objects.

From Earth, a constellation looks like spots of light arranged in a particular shape against the dark night sky. **Figure 1** shows how the constellation of the mythological Greek hunter Orion appears from Earth. It also shows that the stars in a constellation often have no relationship to each other in space.

Stars in the sky can be found at specific locations within a constellation. For example, you can find the star Betelgeuse (BEE tul jooz) in the shoulder of the mighty hunter Orion. Orion's faithful companion is his dog, Canis Major. Sirius, the brightest star that is visible from the northern hemisphere, is in Canis Major.

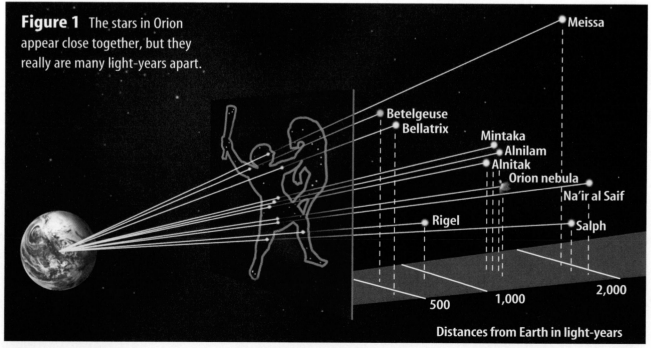

Figure 1 The stars in Orion appear close together, but they really are many light-years apart.

Meissa

Betelgeuse
Bellatrix

Mintaka
Alnilam
Alnitak
Orion nebula
Na'ir al Saif

Rigel
Salph

500   1,000   2,000

Distances from Earth in light-years

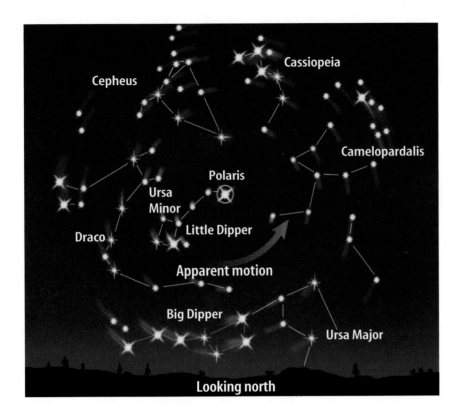

Looking north

Figure 2 The Big Dipper, in red, is part of the constellation Ursa Major. It is visible year-round in the northern hemisphere. Constellations close to Polaris rotate around Polaris, which is almost directly over the north pole.

Modern Constellations Modern astronomy divides the sky into 88 constellations, many of which were named by early astronomers. You probably know some of them. Can you recognize the Big Dipper? It's part of the constellation Ursa Major, shown in **Figure 2.** Notice how the front two stars of the Big Dipper point almost directly at Polaris, which often is called the North Star. Polaris is located at the end of the Little Dipper in the constellation Ursa Minor. It is positioned almost directly over Earth's north pole.

Circumpolar Constellations As Earth rotates, Ursa Major, Ursa Minor, and other constellations in the northern sky circle around Polaris. Because of this, they are called circumpolar constellations. The constellations appear to move, as shown in **Figure 2,** because Earth is in motion. The stars appear to complete one full circle in the sky in about 24 h as Earth rotates on its axis. One circumpolar constellation that's easy to find is Cassiopeia (ka see uh PEE uh). You can look for five bright stars that form a big W or a big M in the northern sky, depending on the season.

As Earth orbits the Sun, different constellations come into view while others disappear. Because of their unique position, circumpolar constellations are visible all year long. Other constellations are not. Orion, which is visible in the winter in the northern hemisphere, can't be seen there in the summer because the daytime side of Earth is facing it.

Mini LAB

Observing Star Patterns

Procedure
1. On a clear night, go outside after dark and study the stars. Take an adult with you.
2. Let your imagination flow to find patterns of stars that look like something familiar.
3. Draw the stars you see, note their positions, and include a drawing of what you think each star pattern resembles.

Analysis
1. Which of your constellations match those observed by your classmates?
2. How can recognizing star patterns be useful?

Try at Home

Absolute and Apparent Magnitudes

When you look at constellations, you'll notice that some stars are brighter than others. For example, Sirius looks much brighter than Rigel. Is Sirius a brighter star, or is it just closer to Earth, making it appear to be brighter? As it turns out, Sirius is 100 times closer to Earth than Rigel is. If Sirius and Rigel were the same distance from Earth, Rigel would appear much brighter in the night sky than Sirius would.

When you refer to the brightness of a star, you can refer to its absolute magnitude or its apparent magnitude. The **absolute magnitude** of a star is a measure of the amount of light it gives off. A measure of the amount of light received on Earth is the **apparent magnitude.** A star that's dim can appear bright in the sky if it's close to Earth, and a star that's bright can appear dim if it's far away. If two stars are the same distance away, what might cause one of them to be brighter than the other?

Reading Check *What is the difference between absolute and apparent magnitude?*

Applying Science

Are distance and brightness related?

The apparent magnitude of a star is affected by its distance from Earth. This activity will help you determine the relationship between distance and brightness.

Identifying the Problem

Luisa conducted an experiment to determine the relationship between distance and the brightness of stars. She used a meterstick, a light meter, and a lightbulb. She placed the bulb at the zero end of the meterstick, then placed the light meter at the 20-cm mark and recorded the distance and the light-meter reading in her data table. Readings are in luxes, which are units for measuring light intensity. Luisa then increased the distance from the bulb to the light meter and took more readings. By examining the data in the table, can you see a relationship between the two variables?

Effect of Distance on Light

Distance (cm)	Meter Reading (luxes)
20	4150.0
40	1037.5
60	461.1
80	259.4

Solving the Problem

1. What happened to the amount of light recorded when the distance was increased from 20 cm to 40 cm? When the distance was increased from 20 cm to 60 cm?
2. What does this indicate about the relationship between light intensity and distance? What would the light intensity be at 100 cm? Would making a graph help you visualize the relationship?

Measurement in Space

How do scientists determine the distance from Earth to nearby stars? One way is to measure parallax—the apparent shift in the position of an object when viewed from two different positions. Extend your arm and look at your thumb first with your left eye closed and then with your right eye closed, as the girl in **Figure 3A** is doing. Your thumb appears to change position with respect to the background. Now do the same experiment with your thumb closer to your face, as shown in **Figure 3B.** What do you observe? The nearer an object is to the observer, the greater its parallax is.

Astronomers can measure the parallax of relatively close stars to determine their distances from Earth. **Figure 4** shows how a close star's position appears to change. Knowing the angle that the star's position changes and the size of Earth's orbit, astronomers can calculate the distance of the star from Earth.

Because space is so vast, a special unit of measure is needed to record distances. Distances between stars and galaxies are measured in light-years. A **light-year** is the distance that light travels in one year. Light travels at 300,000 km/s, or about 9.5 trillion km in one year. The nearest star to Earth, other than the Sun, is Proxima Centauri. Proxima Centauri is a mere 4.3 light-years away, or about 40 trillion km.

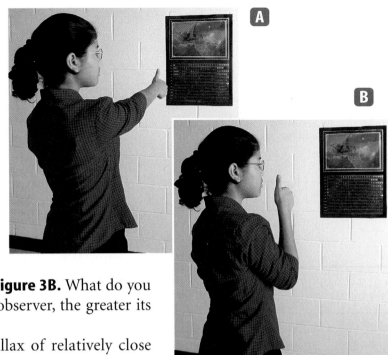

Figure 3 **A** Your thumb appears to move less against the background when it is farther away from your eyes. **B** It appears to move more when it is closer to your eyes.

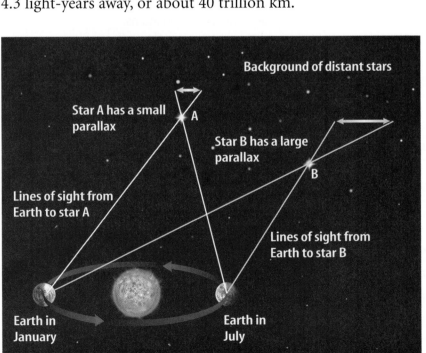

Background of distant stars

Star A has a small parallax

A

Star B has a large parallax

B

Lines of sight from Earth to star A

Lines of sight from Earth to star B

Earth in January

Earth in July

Figure 4 Parallax is determined by observing the same star when Earth is at two different points in its orbit around the Sun. The star's position relative to more distant background stars will appear to change.
Infer *whether star* **A** *or* **B** *is farther from Earth.*

Figure 5 This star spectrum was made by placing a diffraction grating over a telescope's objective lens. A diffraction grating produces a spectrum by causing interference of light waves. **Explain** *what causes the lines in spectra.*

Properties of Stars

The color of a star indicates its temperature. For example, hot stars are a blue-white color. A relatively cool star looks orange or red. Stars that have the same temperature as the Sun have a yellow color.

Astronomers study the composition of stars by observing their spectra. When fitted into a telescope, a spectroscope acts like a prism. It spreads light out in the rainbow band called a spectrum. When light from a star passes through a spectroscope, it breaks into its component colors. Look at the spectrum of a star in **Figure 5.** Notice the dark lines caused by elements in the star's atmosphere. Light radiated from a star passes through the star's atmosphere. As it does, elements in the atmosphere absorb some of this light. The wavelengths of visible light that are absorbed appear as dark lines in the spectrum. Each element absorbs certain wavelengths, producing a unique pattern of dark lines. Like a fingerprint, the patterns of lines can be used to identify the elements in a star's atmosphere.

section 1 review

Summary

Constellations

- Constellations are patterns of stars in the night sky.
- The stars in a constellation often have no relationship to each other in space.

Absolute and Apparent Magnitudes

- Absolute magnitude is a measure of how much light is given off by a star.
- Apparent magnitude is a measure of how much light from a star is received on Earth.

Measurement in Space

- Distances between stars are measured in light-years.

Properties of Stars

- Astronomers study the composition of stars by observing their spectra.

Self Check

1. **Describe** circumpolar constellations.
2. **Explain** why some constellations are visible only during certain seasons.
3. **Infer** how two stars could have the same apparent magnitude but different absolute magnitudes.
4. **Explain** how a star is similar to the Sun if it has the same absorption lines in its spectrum that occur in the Sun's spectrum.
5. **Think Critically** If a star's parallax angle is too small to measure, what can you conclude about the star's distance from Earth?

Applying Skills

6. **Recognize Cause and Effect** Suppose you viewed Proxima Centauri, which is 4.3 light-years from Earth, through a telescope. How old were you when the light that you see left this star?

 Science Online bookj.msscience.com/self_check_quiz

The Sun

The Sun's Layers

The Sun is an ordinary star, but it's important to you. The Sun is the center of the solar system, and the closest star to Earth. Almost all of the life on Earth depends on energy from the Sun.

Notice the different layers of the Sun, shown in **Figure 6,** as you read about them. Like other stars, the Sun is an enormous ball of gas that produces energy by fusing hydrogen into helium in its core. This energy travels outward through the radiation zone and the convection zone. In the convection zone, gases circulate in giant swirls. Finally, energy passes into the Sun's atmosphere.

The Sun's Atmosphere

The lowest layer of the Sun's atmosphere and the layer from which light is given off is the **photosphere.** The photosphere often is called the surface of the Sun, although the surface is not a smooth feature. Temperatures there are about 6,000 K. Above the photosphere is the **chromosphere.** This layer extends upward about 2,000 km above the photosphere. A transition zone occurs between 2,000 km and 10,000 km above the photosphere. Above the transition zone is the **corona.** This is the largest layer of the Sun's atmosphere and extends millions of kilometers into space. Temperatures in the corona are as high as 2 million K. Charged particles continually escape from the corona and move through space as solar wind.

Figure 6 Energy produced in the Sun's core by fusion travels outward by radiation and convection. The Sun's atmosphere shines by the energy produced in the core.

as you read

What You'll Learn

- **Explain** that the Sun is the closest star to Earth.
- **Describe** the structure of the Sun.
- **Describe** sunspots, prominences, and solar flares.

Why It's Important

The Sun is the source of most energy on Earth.

Review Vocabulary

cycle: a repeating sequence of events, such as the sunspot cycle

New Vocabulary

- photosphere
- corona
- chromosphere
- sunspot

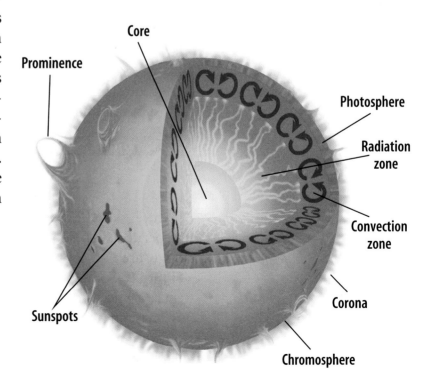

Core

Prominence

Photosphere

Radiation zone

Convection zone

Corona

Chromosphere

Sunspots

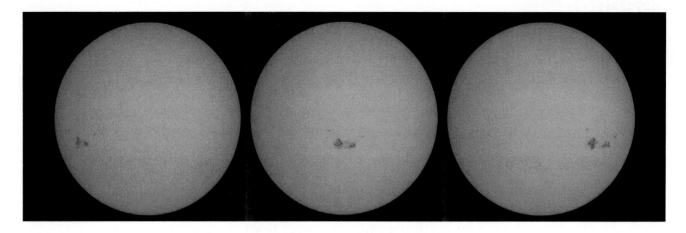

Figure 7 Sunspots are bright, but when viewed against the rest of the photosphere, they appear dark. Notice how these sunspots move as the Sun rotates. **Describe** *the Sun's direction of rotation.*

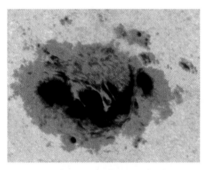

This is a close-up photo of a large sunspot.

Surface Features

From the viewpoint that you observe the Sun, its surface appears to be a smooth layer. But the Sun's surface has many features, including sunspots, prominences, flares, and CMEs.

Sunspots Areas of the Sun's surface that appear dark because they are cooler than surrounding areas are called **sunspots.** Ever since Galileo Galilei made drawings of sunspots, scientists have been studying them. Because scientists could observe the movement of individual sunspots, shown in **Figure 7,** they concluded that the Sun rotates. However, the Sun doesn't rotate as a solid body, as Earth does. It rotates faster at its equator than at its poles. Sunspots at the equator take about 25 days to complete one rotation. Near the poles, they take about 35 days.

Sunspots aren't permanent features on the Sun. They appear and disappear over a period of several days, weeks, or months. The number of sunspots increases and decreases in a fairly regular pattern called the sunspot, or solar activity, cycle. Times when many large sunspots occur are called sunspot maximums. Sunspot maximums occur about every 10 to 11 years. Periods of sunspot minimum occur in between.

Reading Check *What is a sunspot cycle?*

Prominences and Flares Sunspots are related to several features on the Sun's surface. The intense magnetic fields associated with sunspots might cause prominences, which are huge, arching columns of gas. Notice the huge prominence in **Figure 8.** Some prominences blast material from the Sun into space at speeds ranging from 600 km/s to more than 1,000 km/s.

Gases near a sunspot sometimes brighten suddenly, shooting outward at high speed. These violent eruptions are called solar flares. You can see a solar flare in **Figure 8.**

CMEs Coronal mass ejections (CMEs) occur when large amounts of electrically-charged gas are ejected suddenly from the Sun's corona. CMEs can occur as often as two or three times each day during a sunspot maximum.

CMEs present little danger to life on Earth, but they do have some effects. CMEs can damage satellites in orbit around Earth. They also can interfere with radio and power distribution equipment. CMEs often cause auroras. High-energy particles contained in CMEs and the solar wind are carried past Earth's magnetic field. This generates electric currents that flow toward Earth's poles. These electric currents ionize gases in Earth's atmosphere. When these ions recombine with electrons, they produce the light of an aurora, shown in **Figure 8.**

Topic: Space Weather
Visit bookj.msscience.com for Web links to information about space weather and its effects.

Activity Record space weather conditions for several weeks. How does space weather affect Earth?

Figure 8 Features such as solar prominences and solar flares can reach hundreds of thousands of kilometers into space. CMEs are generated as magnetic fields above sunspot groups rearrange. CMEs can trigger events that produce auroras.

Solar prominence

Solar flare

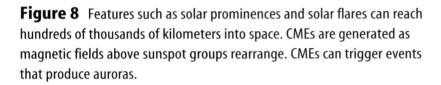

Aurora borealis, or northern lights

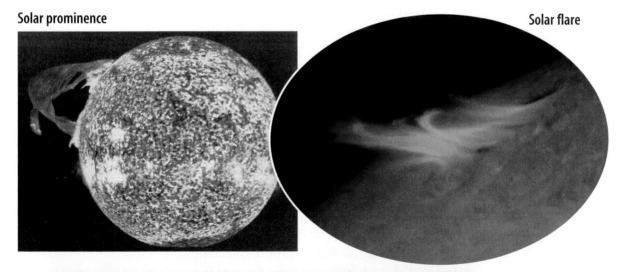

The Sun—An Average Star

The Sun is an average star. It is middle-aged, and its absolute magnitude is about average. It shines with a yellow light. Although the Sun is an average star, it is much closer to Earth than other stars. Light from the Sun reaches Earth in about eight minutes. Light from other stars takes from many years to many millions of years to reach Earth.

The Sun is unusual in one way. It is not close to any other stars. Most stars are part of a system in which two or more stars orbit each other. When two stars orbit each other, it is called a binary system. When three stars orbit each other, it is called a triple star system. The closest star system to the Sun—the Alpha Centauri system, including Proxima Centauri—is a triple star.

Stars also can move through space as a cluster. In a star cluster, many stars are relatively close, so the gravitational attraction among the stars is strong. Most star clusters are far from the solar system. They sometimes appear as a fuzzy patch in the night sky. The double cluster in the northern part of the constellation Perseus is shown in **Figure 9.** On a dark night in autumn, you can see the double cluster with binoculars, but you can't see its individual stars. The Pleiades star cluster can be seen in the constellation of Taurus in the winter sky. On a clear, dark night, you might be able to see seven of the stars in this cluster.

Figure 9 Most stars originally formed in large clusters containing hundreds, or even thousands, of stars.
Draw and label *a sketch of the double cluster.*

---

section 2 review

Summary

The Sun's Layers
- The Sun's interior has layers that include the core, radiation zone, and convection zone.

The Sun's Atmosphere
- The Sun's atmosphere includes the photosphere, chromosphere, and corona.

Surface Features
- The number of sunspots on the Sun varies in a 10- to 11-year cycle.
- Auroras occur when charged particles from the Sun interact with Earth's magnetic field.

The Sun—An Average Star
- The Sun is an average star, but it is much closer to Earth than any other star.

Self Check

1. **Explain** why the Sun is important for life on Earth.
2. **Describe** the sunspot cycle.
3. **Explain** why sunspots appear dark.
4. **Explain** why the Sun, which is an average star, appears so much brighter from Earth than other stars do.
5. **Think Critically** When a CME occurs on the Sun, it takes a couple of days for effects to be noticed on Earth. Explain.

Applying Skills

6. **Communicate** Make a sketch that shows the Sun's layers in your Science Journal. Write a short description of each layer.

Science Online bookj.msscience.com/self_check_quiz

Sunspts

Sunspots can be observed moving across the face of the Sun as it rotates. Measure the movement of sunspots, and use your data to determine the Sun's period of rotation.

● Real-World Question

Can sunspot motion be used to determine the Sun's period of rotation?

Goals

■ **Observe** sunspots and estimate their size.
■ **Estimate** the rate at which sunspots move across the face of the Sun.

Materials

several books   clipboard
piece of cardboard   small tripod
drawing paper   scissors
refracting telescope

Safety Precautions

WARNING: *Handle scissors with care.*

● Procedure

1. Find a location where the Sun can be viewed at the same time of day for a minimum of five days. **WARNING:** *Do not look directly at the Sun. Do not look through the telescope at the Sun. You could damage your eyes.*

2. If the telescope has a small finder scope attached, remove it or keep it covered.

3. Set up the telescope with the eyepiece facing away from the Sun, as shown. Align the telescope so that the shadow it casts on the ground is the smallest size possible. Cut and attach the cardboard as shown in the photo.

4. Use books to prop the clipboard upright. Point the eyepiece at the drawing paper.

5. Move the clipboard back and forth until you have the largest image of the Sun on the paper. Adjust the telescope to form a clear image. Trace the outline of the Sun on the paper.

6. Trace any sunspots that appear as dark areas on the Sun's image. Repeat this step at the same time each day for a week.

7. Using the Sun's diameter (approximately 1,390,000 km), estimate the size of the largest sunspots that you observed.

8. **Calculate** how many kilometers the sunspots move each day.

9. **Predict** how many days it will take for the same group of sunspots to return to the same position in which they appeared on day 1.

● Conclude and Apply

1. What was the estimated size and rate of motion of the largest sunspots?

2. **Infer** how sunspots can be used to determine that the Sun's surface is not solid like Earth's surface.

𝒞ommunicating Your Data

Compare your conclusions with those of other students in your class. **For more help, refer to the** Science Skill Handbook.

Evolution of Stars

as you read

What You'll Learn

- **Describe** how stars are classified.
- **Compare** the Sun to other types of stars on the H-R diagram.
- **Describe** how stars evolve.

Why It's Important

Earth and your body contain elements that were made in stars.

⊙ **Review Vocabulary**
gravity: an attractive force between objects that have mass

New Vocabulary
- nebula
- giant
- white dwarf
- supergiant
- neutron star
- black hole

Classifying Stars

When you look at the night sky, all stars might appear to be similar, but they are quite different. Like people, they vary in age and size, but stars also vary in temperature and brightness.

In the early 1900s, Ejnar Hertzsprung and Henry Russell made some important observations. They noticed that, in general, stars with higher temperatures also have brighter absolute magnitudes.

Hertzsprung and Russell developed a graph, shown in **Figure 10,** to show this relationship. They placed temperatures across the bottom and absolute magnitudes up one side. A graph that shows the relationship of a star's temperature to its absolute magnitude is called a Hertzsprung-Russell (H-R) diagram.

The Main Sequence As you can see, stars seem to fit into specific areas of the graph. Most stars fit into a diagonal band that runs from the upper left to the lower right of the graph. This band, called the main sequence, contains hot, blue, bright stars in the upper left and cool, red, dim stars in the lower right. Yellow main sequence stars, like the Sun, fall in between.

Figure 10 The relationships among a star's color, temperature, and brightness are shown in this H-R diagram. Stars in the upper left are hot, bright stars, and stars in the lower right are cool, dim stars.
Classify *Which type of star shown in the diagram is the hottest, dimmest star?*

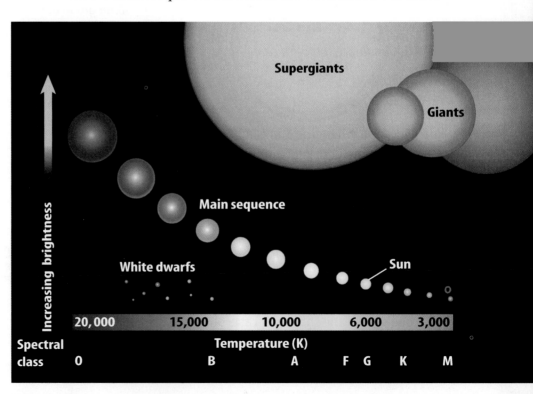

Dwarfs and Giants About 90 percent of all stars are main sequence stars. Most of these are small, red stars found in the lower right of the H-R diagram. Among main sequence stars, the hottest stars generate the most light and the coolest ones generate the least. What about the ten percent of stars that are not part of the main sequence? Some of these stars are hot but not bright. These small stars are located on the lower left of the H-R diagram and are called white dwarfs. Other stars are extremely bright but not hot. These large stars on the upper right of the H-R diagram are called giants, or red giants, because they are usually red in color. The largest giants are called supergiants. **Figure 11** shows the supergiant, Antares—a star 300 times the Sun's diameter—in the constellation Scorpius. It is more than 11,000 times as bright as the Sun.

Reading Check *What kinds of stars are on the main sequence?*

How do stars shine?

For centuries, people were puzzled by the questions of what stars were made of and how they produced light. Many people had estimated that Earth was only a few thousand years old. The Sun could have been made of coal and shined for that long. However, when people realized that Earth was much older, they wondered what material possibly could burn for so many years. Early in the twentieth century, scientists began to understand the process that keeps stars shining for billions of years.

Generating Energy In the 1930s, scientists discovered reactions between the nuclei of atoms. They hypothesized that temperatures in the center of the Sun must be high enough to cause hydrogen to fuse to make helium. This reaction releases tremendous amounts of energy. Much of this energy is emitted as different wavelengths of light, including visible, infrared, and ultraviolet light. Only a tiny fraction of this light comes to Earth. During the fusion reaction, four hydrogen nuclei combine to create one helium nucleus. The mass of one helium nucleus is less than the mass of four hydrogen nuclei, so some mass is lost in the reaction.

Years earlier, in 1905, Albert Einstein had proposed a theory stating that mass can be converted into energy. This was stated as the famous equation $E = mc^2$. In this equation, E is the energy produced, m is the mass, and c is the speed of light. The small amount of mass "lost" when hydrogen atoms fuse to form a helium atom is converted to a large amount of energy.

Figure 11 Antares is a bright supergiant located 400 light-years from Earth. Although its temperature is only about 3,500 K, it is the 16th brightest star in the sky.

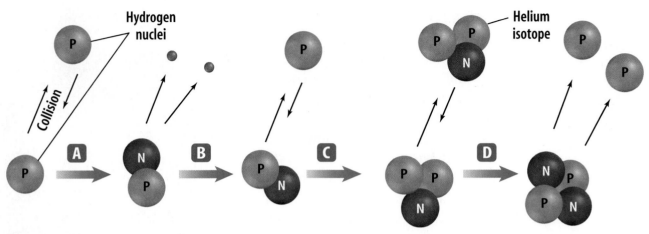

A Two protons (hydrogen nuclei) collide. One proton decays to a neutron, releasing subatomic particles and some energy.

B Another proton fuses with a proton and neutron to form an isotope of helium. Energy is given off again.

C Two helium isotopes collide with enough energy to fuse.

D A helium nucleus (two protons and two neutrons) forms as two protons break away. During the process, still more energy is released.

Figure 12 Fusion of hydrogen into helium occurs in a star's core. **Infer** *what happens to the "lost" mass during this process.*

Fusion Shown in **Figure 12,** fusion occurs in the cores of stars. Only in the core are temperatures high enough to cause atoms to fuse. Normally, they would repel each other, but in the core of a star where temperatures can exceed 15,000,000 K, atoms can move so fast that some of them fuse upon colliding.

Evolution of Stars

 INTEGRATE Physics The H-R diagram explained a lot about stars. However, it also led to more questions. Many wondered why some stars didn't fit in the main sequence group and what happened when a star depleted its supply of hydrogen fuel. Today, scientists have theories about how stars evolve, what makes them different from one another, and what happens when they die. **Figure 13** illustrates the lives of different types of stars.

When hydrogen fuel is depleted, a star loses its main sequence status. This can take less than 1 million years for the brightest stars to many billions of years for the dimmest stars. The Sun has a main sequence life span of about 10 billion years. Half of its life span is still in the future.

Nebula Stars begin as a large cloud of gas and dust called a **nebula.** As the particles of gas and dust exert a gravitational force on each other, the nebula begins to contract. Gravitational forces cause instability within the nebula. The nebula can break apart into smaller and smaller pieces. Each piece eventually might collapse to form a star.

A Star Is Born As the particles in the smaller pieces of nebula move closer together, the temperatures in each nebula piece increase. When the temperature inside the core of a nebula piece reaches 10 million K, fusion begins. The energy released radiates outward through the condensing ball of gas. As the energy radiates into space, stars are born.

Reading Check *How are stars born?*

Main Sequence to Giant Stars In the newly formed star, the heat from fusion causes pressure to increase. This pressure balances the attraction due to gravity. The star becomes a main sequence star. It continues to use its hydrogen fuel.

When hydrogen in the core of the star is depleted, a balance no longer exists between pressure and gravity. The core contracts, and temperatures inside the star increase. This causes the outer layers of the star to expand and cool. In this late stage of its life cycle, a star is called a **giant.**

After the core temperature reaches 100 million K, helium nuclei fuse to form carbon in the giant's core. By this time, the star has expanded to an enormous size, and its outer layers are much cooler than they were when it was a main sequence star. In about 5 billion years, the Sun will become a giant.

White Dwarfs After the star's core uses much of its helium, it contracts even more and its outer layers escape into space. This leaves behind the hot, dense core. At this stage in a star's evolution, it becomes a **white dwarf.** A white dwarf is about the size of Earth. Eventually, the white dwarf will cool and stop giving off light.

INTEGRATE Chemistry

White Dwarf Matter
The matter in white dwarf stars is more than 500,000 times as dense as the matter in Earth. In white dwarf matter, there are free electrons and atomic nuclei. The resistance of the electrons to pack together more provides pressure that keeps the star from collapsing. This state of matter is called electron degeneracy.

Figure 13 The life of a star depends on its mass. Massive stars eventually become neutron stars or black holes.
Explain *what happens to stars that are the size of the Sun.*

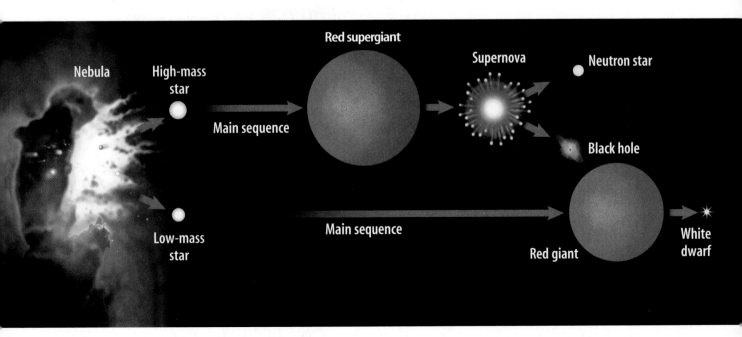

Nebula   High-mass star   Red supergiant   Supernova   Neutron star
Main sequence   Black hole
Low-mass star   Main sequence   Red giant   White dwarf

386 Supernova In 386 A.D., Chinese observers described a new star—a supernova—in the night sky. More recently, astronomers using the *Chandra X-ray Observatory* found evidence of a spinning neutron star, called a pulsar, in exactly the same location. Because of the Chinese account, astronomers better understand how neutron stars form and evolve.

Figure 14 The black hole at the center of galaxy M87 pulls matter into it at extremely high velocities. Some matter is ejected to produce a jet of gas that streams away from the center of the galaxy at nearly light speed.

Supergiants and Supernovas In stars that are more than about eight times more massive than the Sun, the stages of evolution occur more quickly and more violently. Look back at **Figure 13.** In massive stars, the core heats up to much higher temperatures. Heavier and heavier elements form by fusion, and the star expands into a **supergiant.** Eventually, iron forms in the core. Because of iron's atomic structure, it cannot release energy through fusion. The core collapses violently, and a shock wave travels outward through the star. The outer portion of the star explodes, producing a supernova. A supernova can be millions of times brighter than the original star was.

Neutron Stars If the collapsed core of a supernova is between about 1.4 and 3 times as massive as the Sun, it will shrink to approximately 20 km in diameter. Only neutrons can exist in the dense core, and it becomes a **neutron star.** Neutron stars are so dense that a teaspoonful would weigh more than 600 million metric tons in Earth's gravity. As dense as neutron stars are, they can contract only so far because the neutrons resist the inward pull of gravity.

Black Holes If the remaining dense core from a supernova is more than about three times more massive than the Sun, probably nothing can stop the core's collapse. Under these conditions, all of the core's mass collapses to a point. The gravity near this mass is so strong that nothing can escape from it, not even light. Because light cannot escape, the region is called a **black hole.** If you could shine a flashlight on a black hole, the light simply would disappear into it.

✓ **Reading Check** *What is a black hole?*

Black holes, however, are not like giant vacuum cleaners, sucking in distant objects. A black hole has an event horizon, which is a region inside of which nothing can escape. If something—including light—crosses the event horizon, it will be pulled into the black hole. Beyond the event horizon, the black hole's gravity pulls on objects just as it would if the mass had not collapsed. Stars and planets can orbit around a black hole.

The photograph in **Figure 14** was taken by the *Hubble Space Telescope.* It shows a jet of gas streaming out of the center of galaxy M87. This jet of gas formed as matter flowed toward a black hole, and some of the gas was ejected along the polar axis.

Recycling Matter A star begins its life as a nebula, such as the one shown in **Figure 15.** Where does the matter in a nebula come from? Nebulas form partly from the matter that was once in other stars. A star ejects enormous amounts of matter during its lifetime. Some of this matter is incorporated into nebulas, which can evolve to form new stars. The matter in stars is recycled many times.

What about the matter created in the cores of stars and during supernova explosions? Are elements such as carbon and iron also recycled? These elements can become parts of new stars. In fact, spectrographs have shown that the Sun contains some carbon, iron, and other heavier elements. Because the Sun is an average, main sequence star, it is too young and its mass is too small to have formed these elements itself. The Sun condensed from material that was created in stars that died many billions of years ago.

Some elements condense to form planets and other bodies rather than stars. In fact, your body contains many atoms that were fused in the cores of ancient stars. Evidence suggests that the first stars formed from hydrogen and helium and that all the other elements have formed in the cores of stars or as stars explode.

Figure 15 Stars are forming in the Orion Nebula and other similar nebulae.
Describe *a star-forming nebula.*

section 3 review

Summary

Classifying Stars
- Most stars plot on the main sequence of an H-R diagram.
- As stars near the end of their lives, they move off of the main sequence.

How do stars shine?
- Stars shine because of a process called fusion.
- During fusion, nuclei of a lighter element merge to form a heavier element.

Evolution of Stars
- Stars form in regions of gas and dust called nebulae.
- Stars evolve differently depending on how massive they are.

Self Check

1. **Explain** how the Sun is different from other stars on the main sequence. How is it different from a giant star? How is it different from a white dwarf?
2. **Describe** how stars release energy.
3. **Outline** the past and probable future of the Sun.
4. **Define** a black hole.
5. **Think Critically** How can white dwarf stars be both hot and dim?

Applying Math

6. **Convert Units** A neutron star has a diameter of 20 km. One kilometer equals 0.62 miles. What is the neutron star's diameter in miles?

Galaxies and the Universe

Galaxies

If you enjoy science fiction, you might have read about explorers traveling through the galaxy. On their way, they visit planets around other stars and encounter strange alien beings. Although this type of space exploration is futuristic, it is possible to explore galaxies today. Using a variety of telescopes, much is being learned about the Milky Way and other galaxies.

A **galaxy** is a large group of stars, gas, and dust held together by gravity. Earth and the solar system are in a galaxy called the Milky Way. It might contain as many as one trillion stars. Countless other galaxies also exist. Each of these galaxies contains the same elements, forces, and types of energy that occur in Earth's solar system. Galaxies are separated by huge distances—often millions of light-years.

In the same way that stars are grouped together within galaxies, galaxies are grouped into clusters. The cluster that the Milky Way belongs to is called the Local Group. It contains about 45 galaxies of various sizes and types. The three major types of galaxies are spiral, elliptical, and irregular.

Spiral Galaxies Spiral galaxies are galaxies that have spiral arms that wind outward from the center. The arms consist of bright stars, dust, and gas. The Milky Way Galaxy, shown in **Figure 16,** is a spiral galaxy. The Sun and the rest of the solar system are located near the outer edge of the Milky Way Galaxy.

Spiral galaxies can be normal or barred. Arms in a normal spiral start close to the center of the galaxy. Barred spirals have spiral arms extending from a large bar of stars and gas that passes through the center of the galaxy.

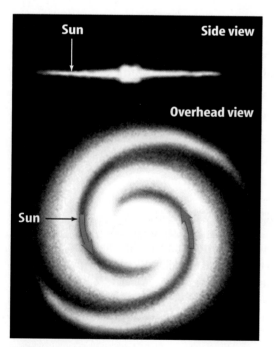

Figure 16 This illustration shows a side view and an overhead view of the Milky Way.
Describe *where the Sun is in the Milky Way.*

Elliptical Galaxies A common type of galaxy is the elliptical galaxy. **Figure 17** shows an elliptical galaxy in the constellation Andromeda. These galaxies are shaped like large, three-dimensional ellipses. Many are football shaped, but others are round. Some elliptical galaxies are small, while others are so large that several galaxies the size of the Milky Way would fit inside one of them.

Irregular Galaxies The third type—an irregular galaxy—includes most of those galaxies that don't fit into the other categories. Irregular galaxies have many different shapes. They are smaller than the other types of galaxies. Two irregular galaxies called the Clouds of Magellan orbit the Milky Way. The Large Magellanic Cloud is shown in **Figure 18.**

 *How do the three different types of galaxies differ?*

The Milky Way Galaxy

The Milky Way might contain one trillion stars. The visible disk of stars shown in **Figure 16** is about 100,000 light-years across. Find the location of the Sun. Notice that it is located about 26,000 light-years from the galaxy's center in one of the spiral arms. In the galaxy, all stars orbit around a central region, or core. It takes about 225 million years for the Sun to orbit the center of the Milky Way.

The Milky Way often is classified as a normal spiral galaxy. However, recent evidence suggests that it might be a barred spiral. It is difficult to know for sure because astronomers have limited data about how the galaxy looks from the outside.

You can't see the shape of the Milky Way because you are located within one of its spiral arms. You can, however, see the Milky Way stretching across the sky as a misty band of faint light. You can see the brightest part of the Milky Way if you look low in the southern sky on a moonless summer night. All the stars you can see in the night sky belong to the Milky Way.

Like many other galaxies, the Milky Way has a supermassive black hole at its center. This black hole might be more than 2.5 million times as massive as the Sun. Evidence for the existence of the black hole comes from observing the orbit of a star near the galaxy's center. Additional evidence includes X-ray emissions detected by the *Chandra X-ray Observatory*. X rays are produced when matter spirals into a black hole.

Figure 17 This photo shows an example of an elliptical galaxy. **Identify** *the two other types of galaxies.*

Figure 18 The Large Magellanic Cloud is an irregular galaxy. It's a member of the Local Group, and it orbits the Milky Way.

Origin of the Universe

People long have wondered how the universe formed. Several models of its origin have been proposed. One model is the steady state theory. It suggests that the universe always has been the same as it is now. The universe always existed and always will. As the universe expands, new matter is created to keep the overall density of the universe the same or in a steady state. However, evidence indicates that the universe was much different in the past.

A second idea is called the oscillating model. In this model, the universe began with expansion. Over time, the expansion slowed and the universe contracted. Then the process began again, oscillating back and forth. Some scientists still hypothesize that the universe expands and contracts in a cycle.

A third model of how the universe formed is called the big bang theory. The universe started with a big bang and has been expanding ever since. This theory will be described later.

Expansion of the Universe

What does it sound like when a train is blowing its whistle while it travels past you? The whistle has a higher pitch as the train approaches you. Then the whistle seems to drop in pitch as the train moves away. This effect is called the Doppler shift. The Doppler shift occurs with light as well as with sound. **Figure 19** shows how the Doppler shift causes changes in the light coming from distant stars and galaxies. If a star is moving toward Earth, its wavelengths of light are compressed. If a star is moving away from Earth, its wavelengths of light are stretched.

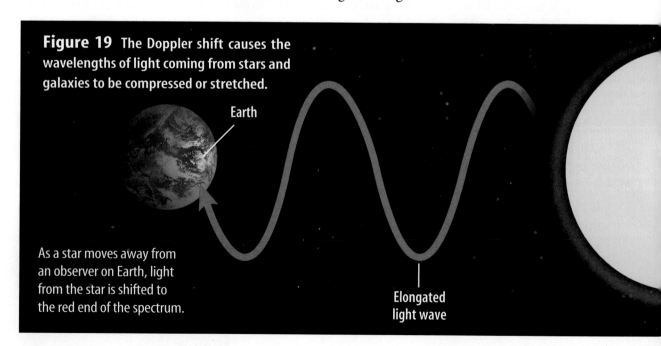

Figure 19 The Doppler shift causes the wavelengths of light coming from stars and galaxies to be compressed or stretched.

Earth

As a star moves away from an observer on Earth, light from the star is shifted to the red end of the spectrum.

Elongated light wave

The Doppler Shift Look at the spectrum of a star in **Figure 20A.** Note the position of the dark lines. How do they compare with the lines in **Figures 20B** and **20C?** They have shifted in position. What caused this shift? As you just read, when a star is moving toward Earth, its wavelengths of light are compressed, just as the sound waves from the train's whistle are. This causes the dark lines in the spectrum to shift toward the blue-violet end of the spectrum. A red shift in the spectrum occurs when a star is moving away from Earth. In a red shift, the dark lines shift toward the red end of the spectrum.

Figure 20 **A** This spectrum shows dark absorption lines. **B** The dark lines shift toward the blue-violet end for a star moving toward Earth. **C** The lines shift toward the red end for a star moving away from Earth.

Red Shift In 1929, Edwin Hubble published an interesting fact about the light coming from most galaxies. When a spectrograph is used to study light from galaxies beyond the Local Group, a red shift occurs in the light. What does this red shift tell you about the universe?

Because all galaxies beyond the Local Group show a red shift in their spectra, they must be moving away from Earth. If all galaxies outside the Local Group are moving away from Earth, then the entire universe must be expanding. Remember the Launch Lab at the beginning of the chapter? The dots on the balloon moved apart as the model universe expanded. Regardless of which dot you picked, all the other dots moved away from it. In a similar way, galaxies beyond the Local Group are moving away from Earth.

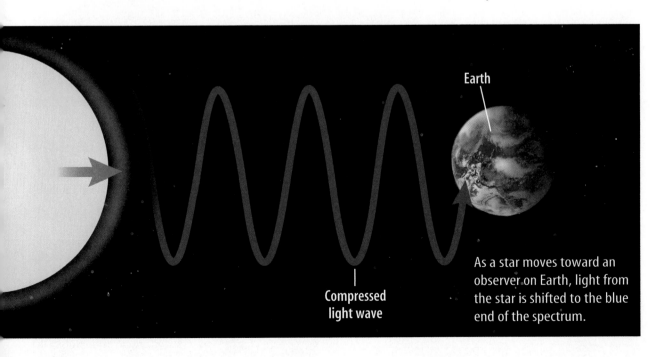

Earth

Compressed light wave

As a star moves toward an observer on Earth, light from the star is shifted to the blue end of the spectrum.

NATIONAL GEOGRAPHIC

VISUALIZING THE BIG BANG THEORY

Figure 21

The big bang theory states that the universe probably began about 13.7 billion years ago with an enormous explosion. Even today, galaxies are rushing apart from this explosion.

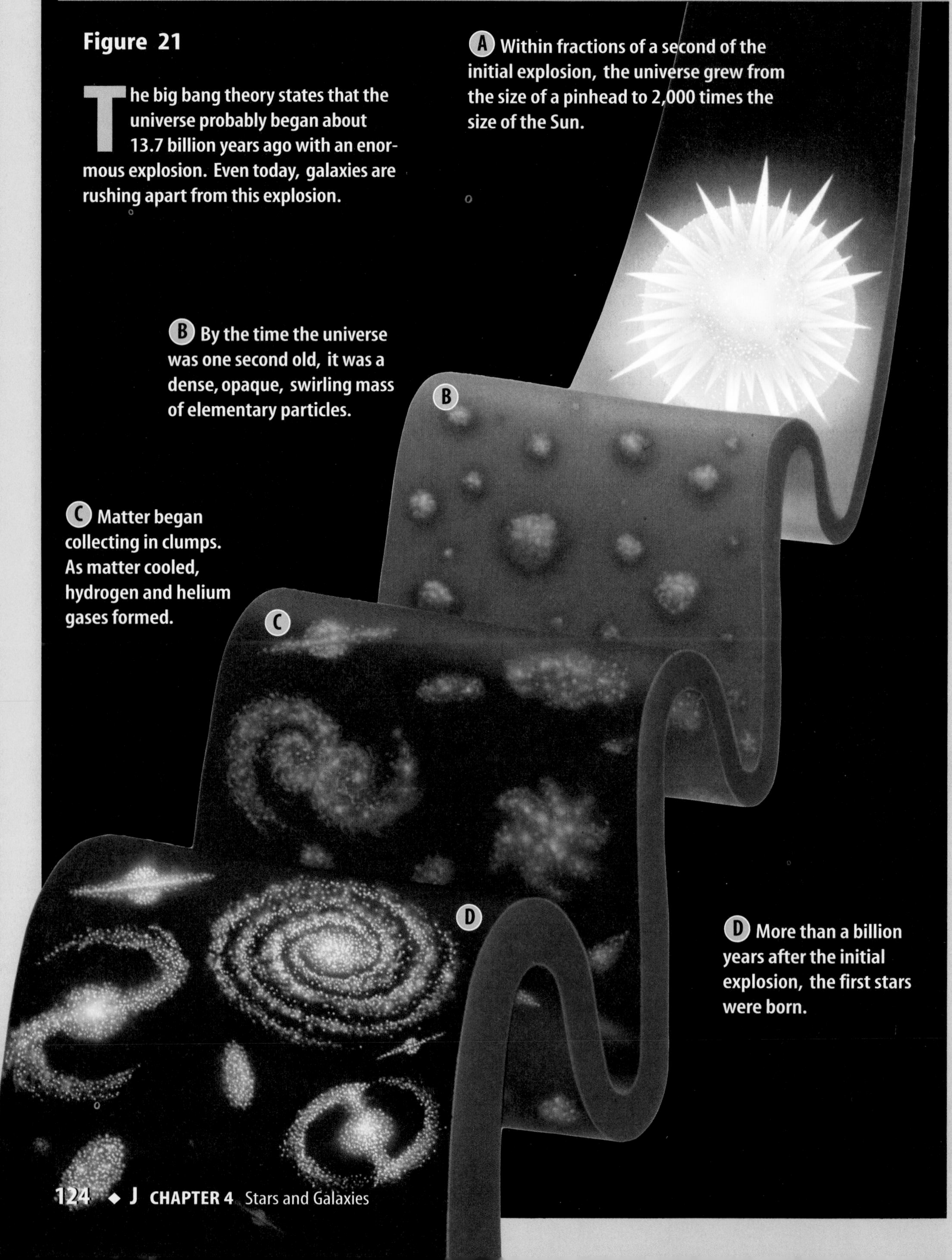

A Within fractions of a second of the initial explosion, the universe grew from the size of a pinhead to 2,000 times the size of the Sun.

B By the time the universe was one second old, it was a dense, opaque, swirling mass of elementary particles.

C Matter began collecting in clumps. As matter cooled, hydrogen and helium gases formed.

D More than a billion years after the initial explosion, the first stars were born.

The Big Bang Theory

When scientists determined that the universe was expanding, they developed a theory to explain their observations. The leading theory about the formation of the universe is called the **big bang theory.** **Figure 21** illustrates the big bang theory. According to this theory, approximately 13.7 billion years ago, the universe began with an enormous explosion. The entire universe began to expand everywhere at the same time.

Looking Back in Time The time-exposure photograph shown in **Figure 22** was taken by the *Hubble Space Telescope*. It shows more than 1,500 galaxies at distances of more than 10 billion light-years. These galaxies could date back to when the universe was no more than 1 billion years old. The galaxies are in various stages of development. One astronomer says that humans might be looking back to a time when the Milky Way was forming.

Whether the universe will expand forever or stop expanding is still unknown. If enough matter exists, gravity might halt the expansion, and the universe will contract until everything comes to a single point. However, studies of distant supernovae indicate that an energy, called dark energy, is causing the universe to expand faster. Scientists are trying to understand how dark energy might affect the fate of the universe.

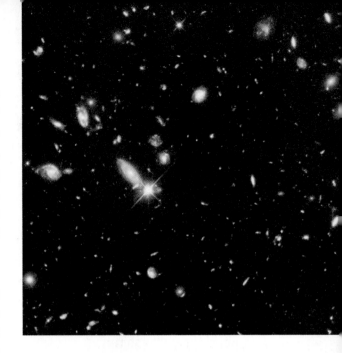

Figure 22 The light from the galaxies in this photo mosaic took billions of years to reach Earth.

section 4 review

Summary

Galaxies
- The three main types of galaxies are spiral, elliptical, and irregular.

The Milky Way Galaxy
- The Milky Way is a spiral galaxy and the Sun is about 26,000 light-years from its center.

Origin of the Universe
- Theories about how the universe formed include the steady state theory, the oscillating universe theory, and the big bang theory.

The Big Bang Theory
- This theory states that the universe began with an explosion about 13.7 billion years ago.

Self Check

1. **Describe** elliptical galaxies. How are they different from spiral galaxies?
2. **Identify** the galaxy that you live in.
3. **Explain** the Doppler shift.
4. **Explain** how all galaxies are similar.
5. **Think Critically** All galaxies outside the Local Group show a red shift. Within the Local Group, some show a red shift and some show a blue shift. What does this tell you about the galaxies in the Local Group?

Applying Skills

6. **Compare and contrast** the theories about the origin of the universe.

Measuring Parallax

Goals

■ **Design** a model to show how the distance from an observer to an object affects the object's parallax shift.

■ **Describe** how parallax can be used to determine the distance to a star.

Possible Materials
meterstick
masking tape
metric ruler
pencil

Safety Precautions

WARNING: *Be sure to wear goggles to protect your eyes.*

◗ Real-World Question

Parallax is the apparent shift in the position of an object when viewed from two locations. How can you build a model to show the relationship between distance and parallax?

◗ Form a Hypothesis

State a hypothesis about how parallax varies with distance.

◗ Test Your Hypothesis

Make a Plan

1. As a group, agree upon and write your hypothesis statement.

2. **List** the steps you need to take to build your model. Be specific, describing exactly what you will do at each step.

3. **Devise** a method to test how distance from an observer to an object, such as a pencil, affects the parallax of the object.

4. **List** the steps you will take to test your hypothesis. Be specific, describing exactly what you will do at each step.

5. Read over your plan for the model to be used in this experiment.

6. How will you determine changes in observed parallax? Remember, these changes should occur when the distance from the observer to the object is changed.

7. You should measure shifts in parallax from several different positions. How will these positions differ?

8. How will you measure distances accurately and compare relative position shift?

Follow Your Plan

1. Make sure your teacher approves your plan before you start.

2. **Construct** the model your team has planned.

3. Carry out the experiment as planned.

4. While conducting the experiment, record any observations that you or other members of your group make in your Science Journal.

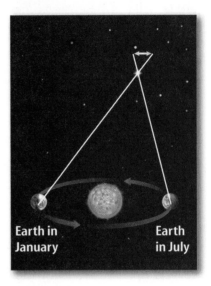

Earth in January

Earth in July

⊙ *Analyze Your Data*

1. **Compare** what happened to the object when it was viewed with one eye closed, then the other.

2. At what distance from the observer did the object appear to shift the most?

3. At what distance did it appear to shift the least?

⊙ *Conclude and Apply*

1. **Infer** what happened to the apparent shift of the object's location as the distance from the observer was increased or decreased.

2. **Describe** how astronomers might use parallax to study stars.

*C*ommunicating
Your Data

Prepare a chart showing the results of your experiment. Share the chart with members of your class. **For more help, refer to the** Science Skill Handbook.

SCIENCE Stats

Stars and Galaxies

Did you know...

. . . A star in Earth's galaxy explodes as a supernova about once a century. The most famous supernova of this galaxy occurred in 1054 and was recorded by the ancient Chinese and Koreans. The explosion was so powerful that it could be seen during the day, and its brightness lasted for weeks. Other major supernovas in the Milky Way that were observed from Earth occurred in 185, 386, 1006, 1181, 1572, and 1604.

Supernova

. . . The large loops of material called solar prominences can extend more than 320,000 km above the Sun's surface. This is so high that two Jupiters and three Earths could fit under the arch.

. . . The red giant star Betelgeuse has a diameter larger than that of Earth's Sun. This gigantic star measures 450,520,000 km in diameter, while the Sun's diameter is a mere 1,390,176 km.

Applying Math Use words to express the number 450,520,000.

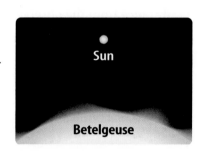
Sun

Betelgeuse

Write About It

Visit bookj.msscience.com/science_stats **to learn whether it might be possible for Earth astronauts to travel to the nearest stars. How long would such a trip take? What problems would have to be overcome? Write a brief report about what you find.**

Reviewing Main Ideas

Section 1 Stars

1. Constellations are patterns of stars in the night sky. Some constellations can be seen all year. Other constellations are visible only during certain seasons.

2. Parallax is the apparent shift in the position of an object when viewed from two different positions. Parallax is used to find the distance to nearby stars.

Section 2 The Sun

1. The Sun is the closest star to Earth.

2. Sunspots are areas on the Sun's surface that are cooler and less bright than surrounding areas.

Section 3 Evolution of Stars

1. Stars are classified according to their position on the H-R diagram.

2. Low-mass stars end their lives as white dwarfs. High-mass stars become neutron stars or black holes.

Section 4 Galaxies and the Universe

1. A galaxy consists of stars, gas, and dust held together by gravity.

2. Earth's solar system is in the Milky Way, a spiral galaxy.

3. The universe is expanding. Scientists don't know whether the universe will expand forever or contract to a single point.

Visualizing Main Ideas

Copy and complete the following concept map that shows the evolution of a main sequence star with a mass similar to that of the Sun.

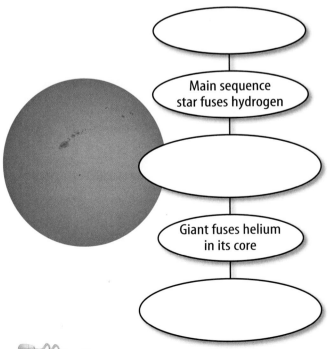

Main sequence
star fuses hydrogen

Giant fuses helium
in its core

Using Vocabulary

absolute magnitude p. 106	giant p. 117
apparent magnitude p. 106	light-year p. 107
big bang theory p. 125	nebula p. 116
black hole p. 118	neutron star p. 118
chromosphere p. 109	photosphere p. 109
constellation p. 104	sunspot p. 110
corona p. 109	supergiant p. 118
galaxy p. 120	white dwarf p. 117

Explain the difference between the terms in each of the following sets.

1. absolute magnitude—apparent magnitude

2. galaxy—constellation

3. giant—supergiant

4. chromosphere—photosphere

5. black hole—neutron star

Checking Concepts

Choose the word or phrase that best answers the question.

6. What is a measure of the amount of a star's light that is received on Earth?
   A) absolute magnitude
   B) apparent magnitude
   C) fusion
   D) parallax

7. What is higher for closer stars?
   A) absolute magnitude
   B) red shift
   C) parallax
   D) blue shift

8. What happens after a nebula contracts and its temperature increases to 10 million K?
   A) a black hole forms
   B) a supernova occurs
   C) fusion begins
   D) a white dwarf forms

9. Which of these has an event horizon?
   A) giant
   B) white dwarf
   C) black hole
   D) neutron star

10. What forms when the Sun fuses hydrogen?
    A) carbon
    B) oxygen
    C) iron
    D) helium

Use the illustration below to answer question 11.

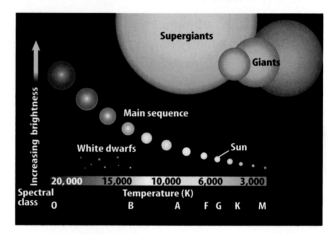

11. Which of the following best describes giant stars?
    A) hot, dim stars
    B) cool, dim stars
    C) hot, bright stars
    D) cool, bright stars

12. Which of the following are loops of matter flowing from the Sun?
    A) sunspots
    B) auroras
    C) coronas
    D) prominences

13. What are groups of galaxies called?
    A) clusters
    B) supergiants
    C) giants
    D) binary systems

14. Which galaxies are sometimes shaped like footballs?
    A) spiral
    B) elliptical
    C) barred
    D) irregular

15. What do scientists study to determine shifts in wavelengths of light?
    A) spectrum
    B) parallax
    C) corona
    D) nebula

Thinking Critically

Use the table below to answer question 16.

Magnitude and Distance of Stars			
Star	Apparent Magnitude	Absolute Magnitude	Distance in Light-Years
A	−26	4.8	0.00002
B	−1.5	1.4	8.7
C	0.1	4.4	4.3
D	0.1	−7.0	815
E	0.4	−5.9	520
F	1.0	−0.6	45

16. Interpret Data Use the table above to answer the following questions. *Hint: lower magnitude values are brighter than higher magnitude values.*
  a. Which star appears brightest from Earth?
  b. Which star would appear brightest from a distance of 10 light-years?
  c. Infer which star in the table above is the Sun.

17. Infer How do scientists know that black holes exist if these objects don't emit visible light?

18. Recognize Cause and Effect Why can parallax only be used to measure distances to stars that are relatively close to Earth?

19. Compare and contrast the Sun with other stars on the H-R diagram.

20. Concept Map Make a concept map showing the life history of a very large star.

21. Make Models Make a model of the Sun. Include all of the Sun's layers in your model.

Performance Activities

22. Story Write a short science-fiction story about an astronaut traveling through the universe. In your story, describe what the astronaut observes. Use as many vocabulary words as you can.

23. Photomontage Gather photographs of the aurora borealis from magazines and other sources. Use the photographs to create a photomontage. Write a caption for each photo.

Applying Math

24. Travel to Vega Vega is a star that is 26 light-years away. If a spaceship could travel at one-tenth the speed of light, how long would it take to reach this star?

Use the illustration below to answer question 25.

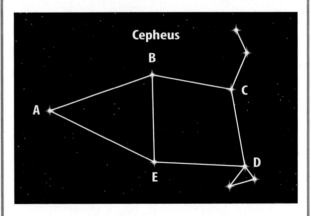

25. Constellation Cepheus The illustration above shows the constellation Cepheus. Answer the following questions about this contellation.
  a. Which of the line segments are nearly parallel?
  b. Which line segments are nearly perpendicular?
  c. Which angles are oblique?
  d. What geometric shape do the three stars at the left side of the drawing form?

Part 1 | Multiple Choice

Record your answers on the answer sheet provided by your teacher or on a sheet of paper.

Use the illustration below to answer question 1.

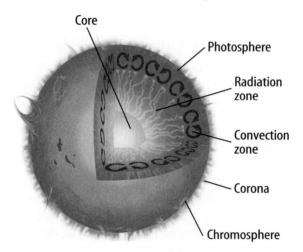

Core

Photosphere

Radiation zone

Convection zone

Corona

Chromosphere

1. The illustration above shows the interior of which object?
   A. Earth   **C.** the Sun
   B. Saturn   **D.** the Moon

2. Which is a group of stars, gas, and dust held together by gravity?
   A. constellation   **C.** black hole
   B. supergiant   **D.** galaxy

3. The most massive stars end their lives as which type of object?
   A. black hole   **C.** neutron star
   B. white dwarf   **D.** black dwarf

4. In which galaxy does the Sun exist?
   A. Arp's Galaxy   **C.** Milky Way Galaxy
   B. Barnard's Galaxy **D.** Andromeda Galaxy

Test-Taking Tip

Process of Elimination If you don't know the answer to a multiple-choice question, eliminate as many incorrect choices as possible. Mark your best guess from the remaining answers before moving on to the next question.

5. Which is the closest star to Earth?
   A. Sirius   **C.** Betelgeuse
   B. the Sun   **D.** the Moon

6. In which of the following choices are the objects ordered from smallest to largest?
   A. stars, galaxies, galaxy clusters, universe
   B. galaxy clusters, galaxies, stars, universe
   C. universe, galaxy clusters, galaxies, stars
   D. universe, stars, galaxies, galaxy clusters

Use the graph below to answer questions 7 and 8.

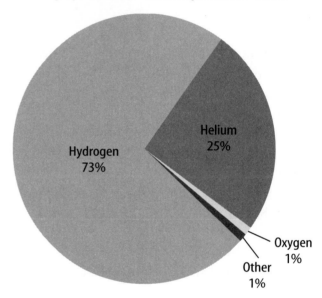

Helium 25%

Hydrogen 73%

Oxygen 1%

Other 1%

7. Which is the most abundant element in the Sun?
   A. helium
   B. hydrogen
   C. oxygen
   D. carbon

8. How will this circle graph change as the Sun ages?
   A. The hydrogen slice will get smaller.
   B. The hydrogen slice will get larger.
   C. The helium slice will get smaller.
   D. The circle graph will not change.

Part 2 | Short Response/Grid In

*Record your answers on the answer sheet
provided by your teacher or on a sheet of paper.*

9. How can events on the Sun affect Earth?
Give one example.

10. How does a red shift differ from a blue
shift?

11. How do astronomers know that the universe is expanding?

12. What is the main sequence?

13. What is a constellation?

Use the illustration below to answer questions 14–16.

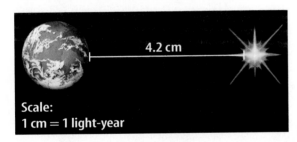

14. According to the illustration, how many
light-years from Earth is Proxima Centauri?

15. How many years would it take for light
from Proxima Centauri to get to Earth?

16. At this scale, how many centimeters would
represent the distance to a star that is 100
light-years from Earth?

17. How can a star's color provide information about its temperature?

18. Approximately how long does it take light
from the Sun to reach Earth? In general,
how does this compare to the amount of
time it takes light from all other stars to
reach Earth?

19. How does the size, temperature, age, and
brightness of the Sun compare to other
stars in the Milky Way Galaxy?

Part 3 | Open Ended

Record your answers on a sheet of paper.

Use the graph below to answer question 20.

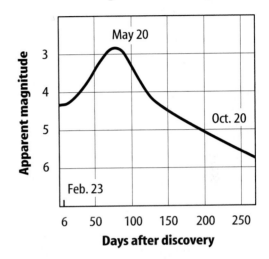

20. The graph above shows the brightness of a
supernova that was observed from Earth
in 1987. Describe how the brightness of
this supernova changed through time.
When was it brightest? What happened
before May 20? What happened after May
20? How much did the brightness change?

21. Compare and contrast the different types
of galaxies.

22. Write a detailed description of the Sun.
What is it? What is it like?

23. Explain how parallax is used to measure
the distance to nearby stars.

24. Why are some constellations visible all
year? Why are other constellations only
visible during certain seasons?

25. What are black holes? How do they form?

26. Explain the big bang theory.

27. What can be learned by studying the dark
lines in a star's spectrum?

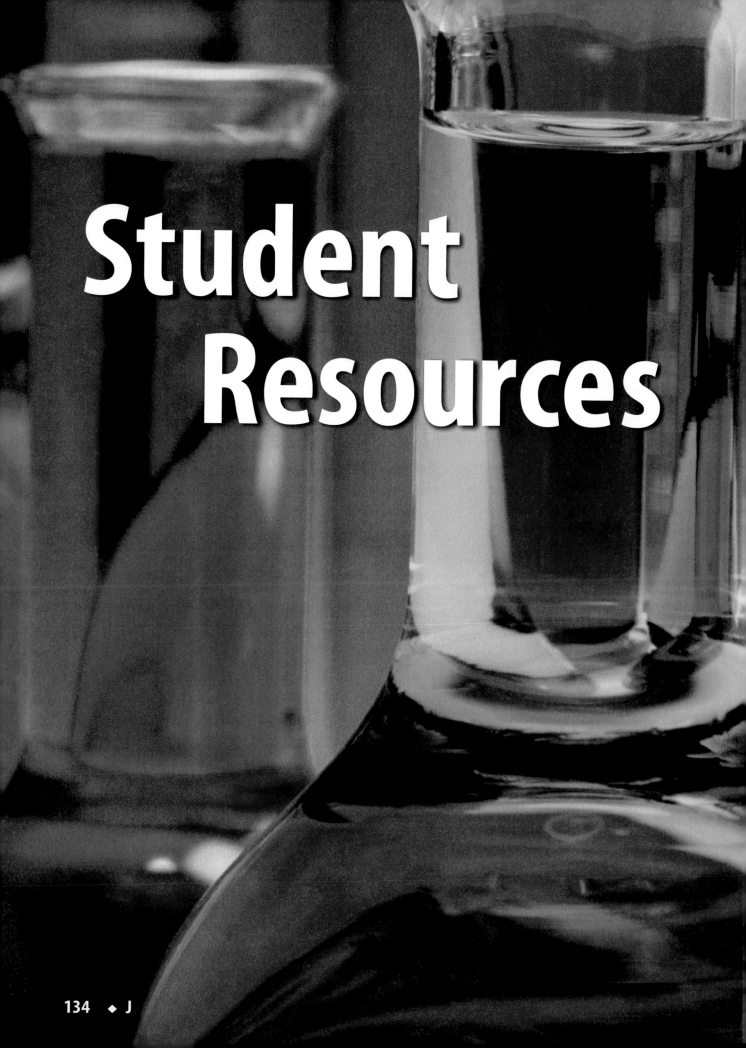

Student Resources

CONTENTS

Scientific Methods

Scientists use an orderly approach called the scientific method to solve problems. This includes organizing and recording data so others can understand them. Scientists use many variations in this method when they solve problems.

Identify a Question

The first step in a scientific investigation or experiment is to identify a question to be answered or a problem to be solved. For example, you might ask which gasoline is the most efficient.

Gather and Organize Information

After you have identified your question, begin gathering and organizing information. There are many ways to gather information, such as researching in a library, interviewing those knowledgeable about the subject, testing and working in the laboratory and field. Fieldwork is investigations and observations done outside of a laboratory.

Researching Information Before moving in a new direction, it is important to gather the information that already is known about the subject. Start by asking yourself questions to determine exactly what you need to know. Then you will look for the information in various reference sources, like the student is doing in **Figure 1.** Some sources may include textbooks, encyclopedias, government documents, professional journals, science magazines, and the Internet. Always list the sources of your information.

Figure 1 The Internet can be a valuable research tool.

Evaluate Sources of Information Not all sources of information are reliable. You should evaluate all of your sources of information, and use only those you know to be dependable. For example, if you are researching ways to make homes more energy efficient, a site written by the U.S. Department of Energy would be more reliable than a site written by a company that is trying to sell a new type of weatherproofing material. Also, remember that research always is changing. Consult the most current resources available to you. For example, a 1985 resource about saving energy would not reflect the most recent findings.

Sometimes scientists use data that they did not collect themselves, or conclusions drawn by other researchers. This data must be evaluated carefully. Ask questions about how the data were obtained, if the investigation was carried out properly, and if it has been duplicated exactly with the same results. Would you reach the same conclusion from the data? Only when you have confidence in the data can you believe it is true and feel comfortable using it.

Interpret Scientific Illustrations As you research a topic in science, you will see drawings, diagrams, and photographs to help you understand what you read. Some illustrations are included to help you understand an idea that you can't see easily by yourself, like the tiny particles in an atom in **Figure 2.** A drawing helps many people to remember details more easily and provides examples that clarify difficult concepts or give additional information about the topic you are studying. Most illustrations have labels or a caption to identify or to provide more information.

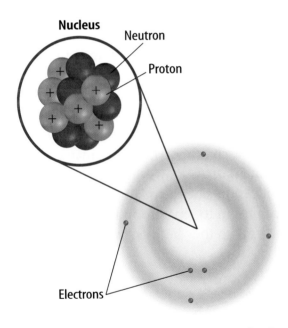

Figure 2 This drawing shows an atom of carbon with its six protons, six neutrons, and six electrons.

Concept Maps One way to organize data is to draw a diagram that shows relationships among ideas (or concepts). A concept map can help make the meanings of ideas and terms more clear, and help you understand and remember what you are studying. Concept maps are useful for breaking large concepts down into smaller parts, making learning easier.

Network Tree A type of concept map that not only shows a relationship, but how the concepts are related is a network tree, shown in **Figure 3.** In a network tree, the words are written in the ovals, while the description of the type of relationship is written across the connecting lines.

When constructing a network tree, write down the topic and all major topics on separate pieces of paper or notecards. Then arrange them in order from general to specific. Branch the related concepts from the major concept and describe the relationship on the connecting line. Continue to more specific concepts until finished.

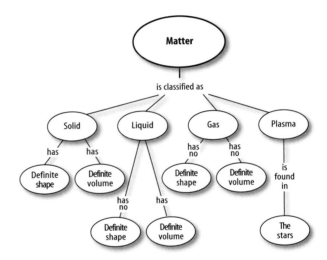

Figure 3 A network tree shows how concepts or objects are related.

Events Chain Another type of concept map is an events chain. Sometimes called a flow chart, it models the order or sequence of items. An events chain can be used to describe a sequence of events, the steps in a procedure, or the stages of a process.

When making an events chain, first find the one event that starts the chain. This event is called the initiating event. Then, find the next event and continue until the outcome is reached, as shown in **Figure 4.**

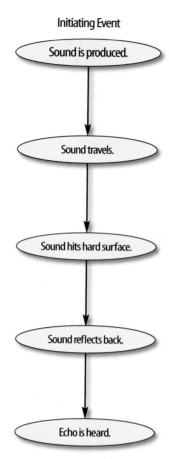

Initiating Event

Figure 4 Events-chain concept maps show the order of steps in a process or event. This concept map shows how a sound makes an echo.

Cycle Map A specific type of events chain is a cycle map. It is used when the series of events do not produce a final outcome, but instead relate back to the beginning event, such as in **Figure 5.** Therefore, the cycle repeats itself.

To make a cycle map, first decide what event is the beginning event. This is also called the initiating event. Then list the next events in the order that they occur, with the last event relating back to the initiating event. Words can be written between the events that describe what happens from one event to the next. The number of events in a cycle map can vary, but usually contain three or more events.

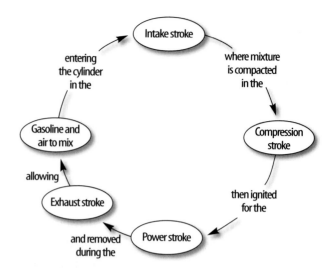

Figure 5 A cycle map shows events that occur in a cycle.

Spider Map A type of concept map that you can use for brainstorming is the spider map. When you have a central idea, you might find that you have a jumble of ideas that relate to it but are not necessarily clearly related to each other. The spider map on sound in **Figure 6** shows that if you write these ideas outside the main concept, then you can begin to separate and group unrelated terms so they become more useful.

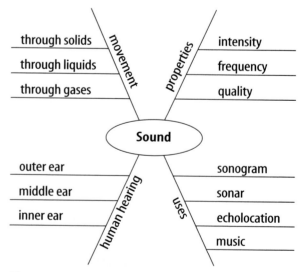

Figure 6 A spider map allows you to list ideas that relate to a central topic but not necessarily to one another.

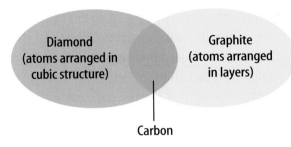

Figure 7 This Venn diagram compares and contrasts two substances made from carbon.

Venn Diagram To illustrate how two subjects compare and contrast you can use a Venn diagram. You can see the characteristics that the subjects have in common and those that they do not, shown in **Figure 7.**

To create a Venn diagram, draw two overlapping ovals that that are big enough to write in. List the characteristics unique to one subject in one oval, and the characteristics of the other subject in the other oval. The characteristics in common are listed in the overlapping section.

Make and Use Tables One way to organize information so it is easier to understand is to use a table. Tables can contain numbers, words, or both.

To make a table, list the items to be compared in the first column and the characteristics to be compared in the first row. The title should clearly indicate the content of the table, and the column or row heads should be clear. Notice that in **Table 1** the units are included.

Table 1 Recyclables Collected During Week			
Day of Week	**Paper (kg)**	**Aluminum (kg)**	**Glass (kg)**
Monday	5.0	4.0	12.0
Wednesday	4.0	1.0	10.0
Friday	2.5	2.0	10.0

Make a Model One way to help you better understand the parts of a structure, the way a process works, or to show things too large or small for viewing is to make a model. For example, an atomic model made of a plastic-ball nucleus and pipe-cleaner electron shells can help you visualize how the parts of an atom relate to each other. Other types of models can by devised on a computer or represented by equations.

Form a Hypothesis

A possible explanation based on previous knowledge and observations is called a hypothesis. After researching gasoline types and recalling previous experiences in your family's car you form a hypothesis—our car runs more efficiently because we use premium gasoline. To be valid, a hypothesis has to be something you can test by using an investigation.

Predict When you apply a hypothesis to a specific situation, you predict something about that situation. A prediction makes a statement in advance, based on prior observation, experience, or scientific reasoning. People use predictions to make everyday decisions. Scientists test predictions by performing investigations. Based on previous observations and experiences, you might form a prediction that cars are more efficient with premium gasoline. The prediction can be tested in an investigation.

Design an Experiment A scientist needs to make many decisions before beginning an investigation. Some of these include: how to carry out the investigation, what steps to follow, how to record the data, and how the investigation will answer the question. It also is important to address any safety concerns.

Test the Hypothesis

Now that you have formed your hypothesis, you need to test it. Using an investigation, you will make observations and collect data, or information. This data might either support or not support your hypothesis. Scientists collect and organize data as numbers and descriptions.

Follow a Procedure In order to know what materials to use, as well as how and in what order to use them, you must follow a procedure. **Figure 8** shows a procedure you might follow to test your hypothesis.

Procedure
1. Use regular gasoline for two weeks.
2. Record the number of kilometers between fill-ups and the amount of gasoline used.
3. Switch to premium gasoline for two weeks.
4. Record the number of kilometers between fill-ups and the amount of gasoline used.

Figure 8 A procedure tells you what to do step by step.

Identify and Manipulate Variables and Controls In any experiment, it is important to keep everything the same except for the item you are testing. The one factor you change is called the independent variable. The change that results is the dependent variable. Make sure you have only one independent variable, to assure yourself of the cause of the changes you observe in the dependent variable. For example, in your gasoline experiment the type of fuel is the independent variable. The dependent variable is the efficiency.

Many experiments also have a control—an individual instance or experimental subject for which the independent variable is not changed. You can then compare the test results to the control results. To design a control you can have two cars of the same type. The control car uses regular gasoline for four weeks. After you are done with the test, you can compare the experimental results to the control results.

Collect Data

Whether you are carrying out an investigation or a short observational experiment, you will collect data, as shown in **Figure 9.** Scientists collect data as numbers and descriptions and organize it in specific ways.

Observe Scientists observe items and events, then record what they see. When they use only words to describe an observation, it is called qualitative data. Scientists' observations also can describe how much there is of something. These observations use numbers, as well as words, in the description and are called quantitative data. For example, if a sample of the element gold is described as being "shiny and very dense" the data are qualitative. Quantitative data on this sample of gold might include "a mass of 30 g and a density of 19.3 g/cm^3."

Figure 9 Collecting data is one way to gather information directly.

Figure 10 Record data neatly and clearly so it is easy to understand.

When you make observations you should examine the entire object or situation first, and then look carefully for details. It is important to record observations accurately and completely. Always record your notes immediately as you make them, so you do not miss details or make a mistake when recording results from memory. Never put unidentified observations on scraps of paper. Instead they should be recorded in a notebook, like the one in **Figure 10.** Write your data neatly so you can easily read it later. At each point in the experiment, record your observations and label them. That way, you will not have to determine what the figures mean when you look at your notes later. Set up any tables that you will need to use ahead of time, so you can record any observations right away. Remember to avoid bias when collecting data by not including personal thoughts when you record observations. Record only what you observe.

Estimate Scientific work also involves estimating. To estimate is to make a judgment about the size or the number of something without measuring or counting. This is important when the number or size of an object or population is too large or too difficult to accurately count or measure.

Sample Scientists may use a sample or a portion of the total number as a type of estimation. To sample is to take a small, representative portion of the objects or organisms of a population for research. By making careful observations or manipulating variables within that portion of the group, information is discovered and conclusions are drawn that might apply to the whole population. A poorly chosen sample can be unrepresentative of the whole. If you were trying to determine the rainfall in an area, it would not be best to take a rainfall sample from under a tree.

Measure You use measurements everyday. Scientists also take measurements when collecting data. When taking measurements, it is important to know how to use measuring tools properly. Accuracy also is important.

Length To measure length, the distance between two points, scientists use meters. Smaller measurements might be measured in centimeters or millimeters.

Length is measured using a metric ruler or meter stick. When using a metric ruler, line up the 0-cm mark with the end of the object being measured and read the number of the unit where the object ends. Look at the metric ruler shown in **Figure 11.** The centimeter lines are the long, numbered lines, and the shorter lines are millimeter lines. In this instance, the length would be 4.50 cm.

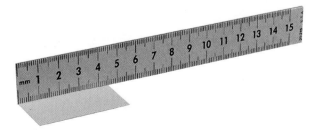

Figure 11 This metric ruler has centimeter and millimeter divisions.

Mass The SI unit for mass is the kilogram (kg). Scientists can measure mass using units formed by adding metric prefixes to the unit gram (g), such as milligram (mg). To measure mass, you might use a triple-beam balance similar to the one shown in **Figure 12.** The balance has a pan on one side and a set of beams on the other side. Each beam has a rider that slides on the beam.

When using a triple-beam balance, place an object on the pan. Slide the largest rider along its beam until the pointer drops below zero. Then move it back one notch. Repeat the process for each rider proceeding from the larger to smaller until the pointer swings an equal distance above and below the zero point. Sum the masses on each beam to find the mass of the object. Move all riders back to zero when finished.

Instead of putting materials directly on the balance, scientists often take a tare of a container. A tare is the mass of a container into which objects or substances are placed for measuring their masses. To mass objects or substances, find the mass of a clean container. Remove the container from the pan, and place the object or substances in the container. Find the mass of the container with the materials in it. Subtract the mass of the empty container from the mass of the filled container to find the mass of the materials you are using.

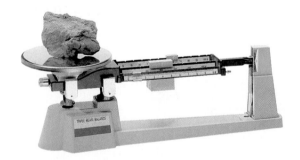

Figure 12 A triple-beam balance is used to determine the mass of an object.

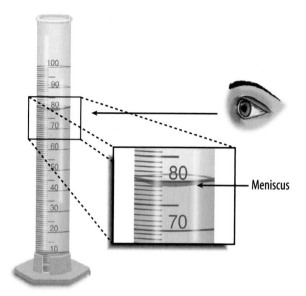

Figure 13 Graduated cylinders measure liquid volume.

Liquid Volume To measure liquids, the unit used is the liter. When a smaller unit is needed, scientists might use a milliliter. Because a milliliter takes up the volume of a cube measuring 1 cm on each side it also can be called a cubic centimeter ($cm^3 = cm \times cm \times cm$).

You can use beakers and graduated cylinders to measure liquid volume. A graduated cylinder, shown in **Figure 13,** is marked from bottom to top in milliliters. In lab, you might use a 10-mL graduated cylinder or a 100-mL graduated cylinder. When measuring liquids, notice that the liquid has a curved surface. Look at the surface at eye level, and measure the bottom of the curve. This is called the meniscus. The graduated cylinder in **Figure 13** contains 79.0 mL, or 79.0 cm^3, of a liquid.

Temperature Scientists often measure temperature using the Celsius scale. Pure water has a freezing point of 0°C and boiling point of 100°C. The unit of measurement is degrees Celsius. Two other scales often used are the Fahrenheit and Kelvin scales.

Figure 14 A thermometer measures the temperature of an object.

Scientists use a thermometer to measure temperature. Most thermometers in a laboratory are glass tubes with a bulb at the bottom end containing a liquid such as colored alcohol. The liquid rises or falls with a change in temperature. To read a glass thermometer like the thermometer in **Figure 14,** rotate it slowly until a red line appears. Read the temperature where the red line ends.

Form Operational Definitions An operational definition defines an object by how it functions, works, or behaves. For example, when you are playing hide and seek and a tree is home base, you have created an operational definition for a tree.

Objects can have more than one operational definition. For example, a ruler can be defined as a tool that measures the length of an object (how it is used). It can also be a tool with a series of marks used as a standard when measuring (how it works).

Analyze the Data

To determine the meaning of your observations and investigation results, you will need to look for patterns in the data. Then you must think critically to determine what the data mean. Scientists use several approaches when they analyze the data they have collected and recorded. Each approach is useful for identifying specific patterns.

Interpret Data The word *interpret* means "to explain the meaning of something." When analyzing data from an experiment, try to find out what the data show. Identify the control group and the test group to see whether or not changes in the independent variable have had an effect. Look for differences in the dependent variable between the control and test groups.

Classify Sorting objects or events into groups based on common features is called classifying. When classifying, first observe the objects or events to be classified. Then select one feature that is shared by some members in the group, but not by all. Place those members that share that feature in a subgroup. You can classify members into smaller and smaller subgroups based on characteristics. Remember that when you classify, you are grouping objects or events for a purpose. Keep your purpose in mind as you select the features to form groups and subgroups.

Compare and Contrast Observations can be analyzed by noting the similarities and differences between two more objects or events that you observe. When you look at objects or events to see how they are similar, you are comparing them. Contrasting is looking for differences in objects or events.

Recognize Cause and Effect A cause is a reason for an action or condition. The effect is that action or condition. When two events happen together, it is not necessarily true that one event caused the other. Scientists must design a controlled investigation to recognize the exact cause and effect.

Draw Conclusions

When scientists have analyzed the data they collected, they proceed to draw conclusions about the data. These conclusions are sometimes stated in words similar to the hypothesis that you formed earlier. They may confirm a hypothesis, or lead you to a new hypothesis.

Infer Scientists often make inferences based on their observations. An inference is an attempt to explain observations or to indicate a cause. An inference is not a fact, but a logical conclusion that needs further investigation. For example, you may infer that a fire has caused smoke. Until you investigate, however, you do not know for sure.

Apply When you draw a conclusion, you must apply those conclusions to determine whether the data supports the hypothesis. If your data do not support your hypothesis, it does not mean that the hypothesis is wrong. It means only that the result of the investigation did not support the hypothesis. Maybe the experiment needs to be redesigned, or some of the initial observations on which the hypothesis was based were incomplete or biased. Perhaps more observation or research is needed to refine your hypothesis. A successful investigation does not always come out the way you originally predicted.

Avoid Bias Sometimes a scientific investigation involves making judgments. When you make a judgment, you form an opinion. It is important to be honest and not to allow any expectations of results to bias your judgments. This is important throughout the entire investigation, from researching to collecting data to drawing conclusions.

Communicate

The communication of ideas is an important part of the work of scientists. A discovery that is not reported will not advance the scientific community's understanding or knowledge. Communication among scientists also is important as a way of improving their investigations.

Scientists communicate in many ways, from writing articles in journals and magazines that explain their investigations and experiments, to announcing important discoveries on television and radio. Scientists also share ideas with colleagues on the Internet or present them as lectures, like the student is doing in **Figure 15.**

Figure 15 A student communicates to his peers about his investigation.

SAFETY SYMBOLS

SAFETY SYMBOLS	HAZARD	EXAMPLES	PRECAUTION	REMEDY
DISPOSAL	Special disposal procedures need to be followed.	certain chemicals, living organisms	Do not dispose of these materials in the sink or trash can.	Dispose of wastes as directed by your teacher.
BIOLOGICAL	Organisms or other biological materials that might be harmful to humans	bacteria, fungi, blood, unpreserved tissues, plant materials	Avoid skin contact with these materials. Wear mask or gloves.	Notify your teacher if you suspect contact with material. Wash hands thoroughly.
EXTREME TEMPERATURE	Objects that can burn skin by being too cold or too hot	boiling liquids, hot plates, dry ice, liquid nitrogen	Use proper protection when handling.	Go to your teacher for first aid.
SHARP OBJECT	Use of tools or glassware that can easily puncture or slice skin	razor blades, pins, scalpels, pointed tools, dissecting probes, broken glass	Practice commonsense behavior and follow guidelines for use of the tool.	Go to your teacher for first aid.
FUME	Possible danger to respiratory tract from fumes	ammonia, acetone, nail polish remover, heated sulfur, moth balls	Make sure there is good ventilation. Never smell fumes directly. Wear a mask.	Leave foul area and notify your teacher immediately.
ELECTRICAL	Possible danger from electrical shock or burn	improper grounding, liquid spills, short circuits, exposed wires	Double-check setup with teacher. Check condition of wires and apparatus.	Do not attempt to fix electrical problems. Notify your teacher immediately.
IRRITANT	Substances that can irritate the skin or mucous membranes of the respiratory tract	pollen, moth balls, steel wool, fiberglass, potassium permanganate	Wear dust mask and gloves. Practice extra care when handling these materials.	Go to your teacher for first aid.
CHEMICAL	Chemicals can react with and destroy tissue and other materials	bleaches such as hydrogen peroxide; acids such as sulfuric acid, hydrochloric acid; bases such as ammonia, sodium hydroxide	Wear goggles, gloves, and an apron.	Immediately flush the affected area with water and notify your teacher.
TOXIC	Substance may be poisonous if touched, inhaled, or swallowed.	mercury, many metal compounds, iodine, poinsettia plant parts	Follow your teacher's instructions.	Always wash hands thoroughly after use. Go to your teacher for first aid.
FLAMMABLE	Flammable chemicals may be ignited by open flame, spark, or exposed heat.	alcohol, kerosene, potassium permanganate	Avoid open flames and heat when using flammable chemicals.	Notify your teacher immediately. Use fire safety equipment if applicable.
OPEN FLAME	Open flame in use, may cause fire.	hair, clothing, paper, synthetic materials	Tie back hair and loose clothing. Follow teacher's instruction on lighting and extinguishing flames.	Notify your teacher immediately. Use fire safety equipment if applicable.

 Eye Safety Proper eye protection should be worn at all times by anyone performing or observing science activities.

 Clothing Protection This symbol appears when substances could stain or burn clothing.

 Animal Safety This symbol appears when safety of animals and students must be ensured.

 Handwashing After the lab, wash hands with soap and water before removing goggles.

Safety in the Science Laboratory

The science laboratory is a safe place to work if you follow standard safety procedures. Being responsible for your own safety helps to make the entire laboratory a safer place for everyone. When performing any lab, read and apply the caution statements and safety symbol listed at the beginning of the lab.

General Safety Rules

1. Obtain your teacher's permission to begin all investigations and use laboratory equipment.

2. Study the procedure. Ask your teacher any questions. Be sure you understand safety symbols shown on the page.

3. Notify your teacher about allergies or other health conditions which can affect your participation in a lab.

4. Learn and follow use and safety procedures for your equipment. If unsure, ask your teacher.

5. Never eat, drink, chew gum, apply cosmetics, or do any personal grooming in the lab. Never use lab glassware as food or drink containers. Keep your hands away from your face and mouth.

6. Know the location and proper use of the safety shower, eye wash, fire blanket, and fire alarm.

Prevent Accidents

1. Use the safety equipment provided to you. Goggles and a safety apron should be worn during investigations.

2. Do NOT use hair spray, mousse, or other flammable hair products. Tie back long hair and tie down loose clothing.

3. Do NOT wear sandals or other open-toed shoes in the lab.

4. Remove jewelry on hands and wrists. Loose jewelry, such as chains and long necklaces, should be removed to prevent them from getting caught in equipment.

5. Do not taste any substances or draw any material into a tube with your mouth.

6. Proper behavior is expected in the lab. Practical jokes and fooling around can lead to accidents and injury.

7. Keep your work area uncluttered.

Laboratory Work

1. Collect and carry all equipment and materials to your work area before beginning a lab.

2. Remain in your own work area unless given permission by your teacher to leave it.

3. Dispose of chemicals and other materials as directed by your teacher. Place broken glass and solid substances in the proper containers. Never discard materials in the sink.

4. Clean your work area.

5. Wash your hands with soap and water thoroughly BEFORE removing your goggles.

Emergencies

1. Report any fire, electrical shock, glassware breakage, spill, or injury, no matter how small, to your teacher immediately. Follow his or her instructions.

2. If your clothing should catch fire, STOP, DROP, and ROLL. If possible, smother it with the fire blanket or get under a safety shower. NEVER RUN.

3. If a fire should occur, turn off all gas and leave the room according to established procedures.

4. In most instances, your teacher will clean up spills. Do NOT attempt to clean up spills unless you are given permission and instructions to do so.

5. If chemicals come into contact with your eyes or skin, notify your teacher immediately. Use the eyewash or flush your skin or eyes with large quantities of water.

6. The fire extinguisher and first-aid kit should only be used by your teacher unless it is an extreme emergency and you have been given permission.

7. If someone is injured or becomes ill, only a professional medical provider or someone certified in first aid should perform first-aid procedures.

3. Always slant test tubes away from yourself and others when heating them, adding substances to them, or rinsing them.

4. If instructed to smell a substance in a container, hold the container a short distance away and fan vapors towards your nose.

5. Do NOT substitute other chemicals/substances for those in the materials list unless instructed to do so by your teacher.

6. Do NOT take any materials or chemicals outside of the laboratory.

7. Stay out of storage areas unless instructed to be there and supervised by your teacher.

Laboratory Cleanup

1. Turn off all burners, water, and gas, and disconnect all electrical devices.

2. Clean all pieces of equipment and return all materials to their proper places.

EXTRA  Labs

From Your Kitchen, Junk Drawer, or Yard

1 Space Probe Flights

▶ Real-World Question

How can we compare the distances traveled by space probes to their destinations?

Possible Materials
- polystyrene balls (5)
- toothpicks (5)
- small stick-on labels (5)
- tennis ball
- meterstick

▶ Procedure

1. Write the names *Mariner 2, Pioneer 10, Mariner 10, Viking 1,* and *Voyager 2* on the five labels and stick each label on a toothpick. Stick a labeled toothpick into each of the polystyrene balls to represent these five United States space probes.

2. Place the tennis ball in an open space such as a basketball court or field.

3. Measure a distance of 0.42 m from the tennis ball and place the *Mariner 2* probe in that spot. Place the *Pioneer 10* probe 6.28 m away, the *Mariner 10* probe 0.92 m from the ball, the *Viking 1* probe 0.78 m away, and the *Voyager 2* probe 43.47 m from the tennis ball.

▶ Conclude and Apply

1. Create a timeline showing the year each probe was launched and its destination and relate this information to the distance traveled.

2. Mercury is 58 million km from the sun and Earth is 150 million km. Use this information to calculate the scale used for this activity.

2 Creating Craters

▶ Real-World Question

Why does the Moon have craters?

Possible Materials
- drink mix or powdered baby formula
- black pepper or paprika
- large, deep cooking tray or large bowl
- marbles
- small, round candies
- aquarium gravel
- tweezers
- bag of cotton balls

▶ Procedure

1. Pour a 3-cm layer of powder over the bottom of a large, deep cooking tray.

2. Sprinkle a fine layer of black pepper over the powder.

3. Lay a 2–3 cm layer of cotton over half of the powder.

4. Drop marbles and other small objects into the powder not covered by the cotton. Carefully remove the objects with tweezers and observe the craters and impact patterns they make.

5. Drop objects on to the half of the tray covered by cotton.

6. Remove the objects and cotton and observe the marks made by objects in the powder.

▶ Conclude and Apply

1. Compare the impacts made by the objects in the powder not covered by cotton with the impacts in the powder covered by cotton.

2. Infer why the Moon has many craters on its surface but Earth does not.

Adult supervision required for all labs.

③ Many Moons

Real-World Question
How do the number of moons of the nine planets compare?

Possible Materials
- golf balls (5)
- softballs (4)
- colored construction paper
- hole puncher
- pennies (10)
- quarters (8)
- meterstick

Procedure
1. Lay the golf balls and softballs on the floor in a row to represent the nine planets. The golf balls should represent the terrestrial planets and the softballs the gas planets.
2. Next to the golf ball representing Earth place one quarter. A quarter represents a moon with a diameter greater than 1,000 km. Research which planets have moons this size and place quarters next to them.
3. Use pennies to represent moons with a diameter between 200–1,000 km. Place pennies next to the planets with moons this size.
4. Use a hole punch to punch out holes from colored construction paper. These holes represent moons smaller than 200 km in diameter. Research which planets have moons this size and place the holes next to them.

Conclude and Apply
1. Infer why terrestrial planets have fewer moons than gas planets.
2. Infer why astronomers do not believe all the moons in the solar system have been discovered.

④ Big Stars

Real-World Question
How does the size of Earth compare to the size of stars?

Possible Materials
- metric ruler
- meterstick
- tape measure
- masking tape
- white paper
- black marker

Procedure
1. Tape a sheet of white paper to the floor.
2. Draw a dot in the center to the paper. Measure a 1-mm distance from the dot and draw a second dot. This distance represents the diameter of Earth.
3. Measure a distance of 10.9 cm from the first dot and draw a third dot. This distance represents the diameter of the Sun.
4. Measure a distance of 5 m from the first dot and mark the location on the floor with a piece of masking tape. This distance represents the average diameter of a red giant star.
5. Measure a distance of 30 m from the first dot and mark the location on the floor with a piece of masking tape. This distance represents the diameter of the supergiant star Antares.

Conclude and Apply
1. The diameter of Earth is 12,756 km. What is the diameter of the Sun?
2. What is the diameter of an average red giant?

Computer Skills

People who study science rely on computers, like the one in **Figure 16,** to record and store data and to analyze results from investigations. Whether you work in a laboratory or just need to write a lab report with tables, good computer skills are a necessity.

Using the computer comes with responsibility. Issues of ownership, security, and privacy can arise. Remember, if you did not author the information you are using, you must provide a source for your information. Also, anything on a computer can be accessed by others. Do not put anything on the computer that you would not want everyone to know. To add more security to your work, use a password.

Use a Word Processing Program

A computer program that allows you to type your information, change it as many times as you need to, and then print it out is called a word processing program. Word processing programs also can be used to make tables.

Figure 16 A computer will make reports neater and more professional looking.

Learn the Skill To start your word processing program, a blank document, sometimes called "Document 1," appears on the screen. To begin, start typing. To create a new document, click the *New* button on the standard tool bar. These tips will help you format the document.

- The program will automatically move to the next line; press *Enter* if you wish to start a new paragraph.
- Symbols, called non-printing characters, can be hidden by clicking the *Show/Hide* button on your toolbar.
- To insert text, move the cursor to the point where you want the insertion to go, click on the mouse once, and type the text.
- To move several lines of text, select the text and click the *Cut* button on your toolbar. Then position your cursor in the location that you want to move the cut text and click *Paste.* If you move to the wrong place, click *Undo.*
- The spell check feature does not catch words that are misspelled to look like other words, like "cold" instead of "gold." Always reread your document to catch all spelling mistakes.
- To learn about other word processing methods, read the user's manual or click on the *Help* button.
- You can integrate databases, graphics, and spreadsheets into documents by copying from another program and pasting it into your document, or by using desktop publishing (DTP). DTP software allows you to put text and graphics together to finish your document with a professional look. This software varies in how it is used and its capabilities.

Use a Database

A collection of facts stored in a computer and sorted into different fields is called a database. A database can be reorganized in any way that suits your needs.

Learn the Skill A computer program that allows you to create your own database is a database management system (DBMS). It allows you to add, delete, or change information. Take time to get to know the features of your database software.

- Determine what facts you would like to include and research to collect your information.
- Determine how you want to organize the information.
- Follow the instructions for your particular DBMS to set up fields. Then enter each item of data in the appropriate field.
- Follow the instructions to sort the information in order of importance.
- Evaluate the information in your database, and add, delete, or change as necessary.

Use the Internet

The Internet is a global network of computers where information is stored and shared. To use the Internet, like the students in **Figure 17,** you need a modem to connect your computer to a phone line and an Internet Service Provider account.

Learn the Skill To access internet sites and information, use a "Web browser," which lets you view and explore pages on the World Wide Web. Each page is its own site, and each site has its own address, called a URL. Once you have found a Web browser, follow these steps for a search (this also is how you search a database).

Figure 17 The Internet allows you to search a global network for a variety of information.

- Be as specific as possible. If you know you want to research "gold," don't type in "elements." Keep narrowing your search until you find what you want.
- Web sites that end in *.com* are commercial Web sites; *.org, .edu,* and *.gov* are nonprofit, educational, or government Web sites.
- Electronic encyclopedias, almanacs, indexes, and catalogs will help locate and select relevant information.
- Develop a "home page" with relative ease. When developing a Web site, NEVER post pictures or disclose personal information such as location, names, or phone numbers. Your school or community usually can host your Web site. A basic understanding of HTML (hypertext mark-up language), the language of Web sites, is necessary. Software that creates HTML code is called authoring software, and can be downloaded free from many Web sites. This software allows text and pictures to be arranged as the software is writing the HTML code.

Use a Spreadsheet

A spreadsheet, shown in **Figure 18,** can perform mathematical functions with any data arranged in columns and rows. By entering a simple equation into a cell, the program can perform operations in specific cells, rows, or columns.

Learn the Skill Each column (vertical) is assigned a letter, and each row (horizontal) is assigned a number. Each point where a row and column intersect is called a cell, and is labeled according to where it is located—Column A, Row 1 (A1).

- Decide how to organize the data, and enter it in the correct row or column.
- Spreadsheets can use standard formulas or formulas can be customized to calculate cells.
- To make a change, click on a cell to make it activate, and enter the edited data or formula.
- Spreadsheets also can display your results in graphs. Choose the style of graph that best represents the data.

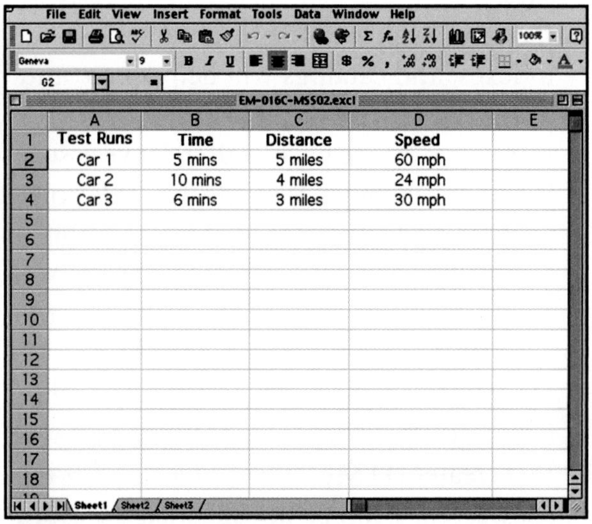

Figure 18 A spreadsheet allows you to perform mathematical operations on your data.

Use Graphics Software

Adding pictures, called graphics, to your documents is one way to make your documents more meaningful and exciting. This software adds, edits, and even constructs graphics. There is a variety of graphics software programs. The tools used for drawing can be a mouse, keyboard, or other specialized devices. Some graphics programs are simple. Others are complicated, called computer-aided design (CAD) software.

Learn the Skill It is important to have an understanding of the graphics software being used before starting. The better the software is understood, the better the results. The graphics can be placed in a word-processing document.

- Clip art can be found on a variety of internet sites, and on CDs. These images can be copied and pasted into your document.
- When beginning, try editing existing drawings, then work up to creating drawings.
- The images are made of tiny rectangles of color called pixels. Each pixel can be altered.
- Digital photography is another way to add images. The photographs in the memory of a digital camera can be downloaded into a computer, then edited and added to the document.
- Graphics software also can allow animation. The software allows drawings to have the appearance of movement by connecting basic drawings automatically. This is called in-betweening, or tweening.
- Remember to save often.

Presentation Skills

Develop Multimedia Presentations

Most presentations are more dynamic if they include diagrams, photographs, videos, or sound recordings, like the one shown in **Figure 19.** A multimedia presentation involves using stereos, overhead projectors, televisions, computers, and more.

Learn the Skill Decide the main points of your presentation, and what types of media would best illustrate those points.

- Make sure you know how to use the equipment you are working with.
- Practice the presentation using the equipment several times.
- Enlist the help of a classmate to push play or turn lights out for you. Be sure to practice your presentation with him or her.
- If possible, set up all of the equipment ahead of time, and make sure everything is working properly.

Figure 19 These students are engaging the audience using a variety of tools.

Computer Presentations

There are many different interactive computer programs that you can use to enhance your presentation. Most computers have a compact disc (CD) drive that can play both CDs and digital video discs (DVDs). Also, there is hardware to connect a regular CD, DVD, or VCR. These tools will enhance your presentation.

Another method of using the computer to aid in your presentation is to develop a slide show using a computer program. This can allow movement of visuals at the presenter's pace, and can allow for visuals to build on one another.

Learn the Skill In order to create multimedia presentations on a computer, you need to have certain tools. These may include traditional graphic tools and drawing programs, animation programs, and authoring systems that tie everything together. Your computer will tell you which tools it supports. The most important step is to learn about the tools that you will be using.

- Often, color and strong images will convey a point better than words alone. Use the best methods available to convey your point.
- As with other presentations, practice many times.
- Practice your presentation with the tools you and any assistants will be using.
- Maintain eye contact with the audience. The purpose of using the computer is not to prompt the presenter, but to help the audience understand the points of the presentation.

Math Review

Use Fractions

A fraction compares a part to a whole. In the fraction $\frac{2}{3}$, the 2 represents the part and is the numerator. The 3 represents the whole and is the denominator.

Reduce Fractions To reduce a fraction, you must find the largest factor that is common to both the numerator and the denominator, the greatest common factor (GCF). Divide both numbers by the GCF. The fraction has then been reduced, or it is in its simplest form.

Example Twelve of the 20 chemicals in the science lab are in powder form. What fraction of the chemicals used in the lab are in powder form?

Step 1 Write the fraction.

$$\frac{\text{part}}{\text{whole}} = \frac{12}{20}$$

Step 2 To find the GCF of the numerator and denominator, list all of the factors of each number.

Factors of 12: 1, 2, 3, 4, 6, 12 (the numbers that divide evenly into 12)

Factors of 20: 1, 2, 4, 5, 10, 20 (the numbers that divide evenly into 20)

Step 3 List the common factors.

1, 2, 4.

Step 4 Choose the greatest factor in the list. The GCF of 12 and 20 is 4.

Step 5 Divide the numerator and denominator by the GCF.

$$\frac{12 \div 4}{20 \div 4} = \frac{3}{5}$$

In the lab, $\frac{3}{5}$ of the chemicals are in powder form.

Practice Problem At an amusement park, 66 of 90 rides have a height restriction. What fraction of the rides, in its simplest form, has a height restriction?

Add and Subtract Fractions To add or subtract fractions with the same denominator, add or subtract the numerators and write the sum or difference over the denominator. After finding the sum or difference, find the simplest form for your fraction.

Example 1 In the forest outside your house, $\frac{1}{8}$ of the animals are rabbits, $\frac{3}{8}$ are squirrels, and the remainder are birds and insects. How many are mammals?

Step 1 Add the numerators.

$$\frac{1}{8} + \frac{3}{8} = \frac{(1+3)}{8} = \frac{4}{8}$$

Step 2 Find the GCF.

$$\frac{4}{8} \quad (\text{GCF, 4})$$

Step 3 Divide the numerator and denominator by the GCF.

$$\frac{4}{4} = 1, \ \frac{8}{4} = 2$$

$\frac{1}{2}$ of the animals are mammals.

Example 2 If $\frac{7}{16}$ of the Earth is covered by freshwater, and $\frac{1}{16}$ of that is in glaciers, how much freshwater is not frozen?

Step 1 Subtract the numerators.

$$\frac{7}{16} - \frac{1}{16} = \frac{(7-1)}{16} = \frac{6}{16}$$

Step 2 Find the GCF.

$$\frac{6}{16} \quad (\text{GCF, 2})$$

Step 3 Divide the numerator and denominator by the GCF.

$$\frac{6}{2} = 3, \ \frac{16}{2} = 8$$

$\frac{3}{8}$ of the freshwater is not frozen.

Practice Problem A bicycle rider is going 15 km/h for $\frac{4}{9}$ of his ride, 10 km/h for $\frac{2}{9}$ of his ride, and 8 km/h for the remainder of the ride. How much of his ride is he going over 8 km/h?

Unlike Denominators To add or subtract fractions with unlike denominators, first find the least common denominator (LCD). This is the smallest number that is a common multiple of both denominators. Rename each fraction with the LCD, and then add or subtract. Find the simplest form if necessary.

Example 1 A chemist makes a paste that is $\frac{1}{2}$ table salt (NaCl), $\frac{1}{3}$ sugar ($C_6H_{12}O_6$), and the rest water (H_2O). How much of the paste is a solid?

Step 1 Find the LCD of the fractions.

$$\frac{1}{2} + \frac{1}{3} \text{ (LCD, 6)}$$

Step 2 Rename each numerator and each denominator with the LCD.

$$1 \times 3 = 3, \quad 2 \times 3 = 6$$
$$1 \times 2 = 2, \quad 3 \times 2 = 6$$

Step 3 Add the numerators.

$$\frac{3}{6} + \frac{2}{6} = \frac{(3 + 2)}{6} = \frac{5}{6}$$

$\frac{5}{6}$ of the paste is a solid.

Example 2 The average precipitation in Grand Junction, CO, is $\frac{7}{10}$ inch in November, and $\frac{3}{5}$ inch in December. What is the total average precipitation?

Step 1 Find the LCD of the fractions.

$$\frac{7}{10} + \frac{3}{5} \text{ (LCD, 10)}$$

Step 2 Rename each numerator and each denominator with the LCD.

$$7 \times 1 = 7, \quad 10 \times 1 = 10$$
$$3 \times 2 = 6, \quad 5 \times 2 = 10$$

Step 3 Add the numerators.

$$\frac{7}{10} + \frac{6}{10} = \frac{(7 + 6)}{10} = \frac{13}{10}$$

$\frac{13}{10}$ inches total precipitation, or $1\frac{3}{10}$ inches.

Practice Problem On an electric bill, about $\frac{1}{8}$ of the energy is from solar energy and about $\frac{1}{10}$ is from wind power. How much of the total bill is from solar energy and wind power combined?

Example 3 In your body, $\frac{7}{10}$ of your muscle contractions are involuntary (cardiac and smooth muscle tissue). Smooth muscle makes $\frac{3}{15}$ of your muscle contractions. How many of your muscle contractions are made by cardiac muscle?

Step 1 Find the LCD of the fractions.

$$\frac{7}{10} - \frac{3}{15} \text{ (LCD, 30)}$$

Step 2 Rename each numerator and each denominator with the LCD.

$$7 \times 3 = 21, \quad 10 \times 3 = 30$$
$$3 \times 2 = 6, \quad 15 \times 2 = 30$$

Step 3 Subtract the numerators.

$$\frac{21}{30} - \frac{6}{30} = \frac{(21 - 6)}{30} = \frac{15}{30}$$

Step 4 Find the GCF.

$$\frac{15}{30} \text{ (GCF, 15)}$$

$$\frac{1}{2}$$

$\frac{1}{2}$ of all muscle contractions are cardiac muscle.

Example 4 Tony wants to make cookies that call for $\frac{3}{4}$ of a cup of flour, but he only has $\frac{1}{3}$ of a cup. How much more flour does he need?

Step 1 Find the LCD of the fractions.

$$\frac{3}{4} - \frac{1}{3} \text{ (LCD, 12)}$$

Step 2 Rename each numerator and each denominator with the LCD.

$$3 \times 3 = 9, \quad 4 \times 3 = 12$$
$$1 \times 4 = 4, \quad 3 \times 4 = 12$$

Step 3 Subtract the numerators.

$$\frac{9}{12} - \frac{4}{12} = \frac{(9 - 4)}{12} = \frac{5}{12}$$

$\frac{5}{12}$ of a cup of flour.

Practice Problem Using the information provided to you in Example 3 above, determine how many muscle contractions are voluntary (skeletal muscle).

Multiply Fractions To multiply with fractions, multiply the numerators and multiply the denominators. Find the simplest form if necessary.

Example Multiply $\frac{3}{5}$ by $\frac{1}{3}$.

Step 1 Multiply the numerators and denominators.
$$\frac{3}{5} \times \frac{1}{3} = \frac{(3 \times 1)}{(5 \times 3)} = \frac{3}{15}$$

Step 2 Find the GCF.
$$\frac{3}{15} \quad (\text{GCF, 3})$$

Step 3 Divide the numerator and denominator by the GCF.
$$\frac{3}{3} = 1, \quad \frac{15}{3} = 5$$
$$\frac{1}{5}$$

$\frac{3}{5}$ multiplied by $\frac{1}{3}$ is $\frac{1}{5}$.

Practice Problem Multiply $\frac{3}{14}$ by $\frac{5}{16}$.

Find a Reciprocal Two numbers whose product is 1 are called multiplicative inverses, or reciprocals.

Example Find the reciprocal of $\frac{3}{8}$.

Step 1 Inverse the fraction by putting the denominator on top and the numerator on the bottom.
$$\frac{8}{3}$$

The reciprocal of $\frac{3}{8}$ is $\frac{8}{3}$.

Practice Problem Find the reciprocal of $\frac{4}{9}$.

Divide Fractions To divide one fraction by another fraction, multiply the dividend by the reciprocal of the divisor. Find the simplest form if necessary.

Example 1 Divide $\frac{1}{9}$ by $\frac{1}{3}$.

Step 1 Find the reciprocal of the divisor.
The reciprocal of $\frac{1}{3}$ is $\frac{3}{1}$.

Step 2 Multiply the dividend by the reciprocal of the divisor.
$$\frac{\frac{1}{9}}{\frac{1}{3}} = \frac{1}{9} \times \frac{3}{1} = \frac{(1 \times 3)}{(9 \times 1)} = \frac{3}{9}$$

Step 3 Find the GCF.
$$\frac{3}{9} \quad (\text{GCF, 3})$$

Step 4 Divide the numerator and denominator by the GCF.
$$\frac{3}{3} = 1, \quad \frac{9}{3} = 3$$
$$\frac{1}{3}$$

$\frac{1}{9}$ divided by $\frac{1}{3}$ is $\frac{1}{3}$.

Example 2 Divide $\frac{3}{5}$ by $\frac{1}{4}$.

Step 1 Find the reciprocal of the divisor.
The reciprocal of $\frac{1}{4}$ is $\frac{4}{1}$.

Step 2 Multiply the dividend by the reciprocal of the divisor.
$$\frac{\frac{3}{5}}{\frac{1}{4}} = \frac{3}{5} \times \frac{4}{1} = \frac{(3 \times 4)}{(5 \times 1)} = \frac{12}{5}$$

$\frac{3}{5}$ divided by $\frac{1}{4}$ is $\frac{12}{5}$ or $2\frac{2}{5}$.

Practice Problem Divide $\frac{3}{11}$ by $\frac{7}{10}$.

Use Ratios

When you compare two numbers by division, you are using a ratio. Ratios can be written 3 to 5, 3:5, or $\frac{3}{5}$. Ratios, like fractions, also can be written in simplest form.

Ratios can represent probabilities, also called odds. This is a ratio that compares the number of ways a certain outcome occurs to the number of outcomes. For example, if you flip a coin 100 times, what are the odds that it will come up heads? There are two possible outcomes, heads or tails, so the odds of coming up heads are 50:100. Another way to say this is that 50 out of 100 times the coin will come up heads. In its simplest form, the ratio is 1:2.

Example 1 A chemical solution contains 40 g of salt and 64 g of baking soda. What is the ratio of salt to baking soda as a fraction in simplest form?

Step 1 Write the ratio as a fraction.
$$\frac{\text{salt}}{\text{baking soda}} = \frac{40}{64}$$

Step 2 Express the fraction in simplest form.
The GCF of 40 and 64 is 8.
$$\frac{40}{64} = \frac{40 \div 8}{64 \div 8} = \frac{5}{8}$$

The ratio of salt to baking soda in the sample is 5:8.

Example 2 Sean rolls a 6-sided die 6 times. What are the odds that the side with a 3 will show?

Step 1 Write the ratio as a fraction.
$$\frac{\text{number of sides with a 3}}{\text{number of sides}} = \frac{1}{6}$$

Step 2 Multiply by the number of attempts.
$$\frac{1}{6} \times 6 \text{ attempts} = \frac{6}{6} \text{ attempts} = 1 \text{ attempt}$$

1 attempt out of 6 will show a 3.

Practice Problem Two metal rods measure 100 cm and 144 cm in length. What is the ratio of their lengths in simplest form?

Use Decimals

A fraction with a denominator that is a power of ten can be written as a decimal. For example, 0.27 means $\frac{27}{100}$. The decimal point separates the ones place from the tenths place.

Any fraction can be written as a decimal using division. For example, the fraction $\frac{5}{8}$ can be written as a decimal by dividing 5 by 8. Written as a decimal, it is 0.625.

Add or Subtract Decimals When adding and subtracting decimals, line up the decimal points before carrying out the operation.

Example 1 Find the sum of 47.68 and 7.80.

Step 1 Line up the decimal places when you write the numbers.
$$47.68$$
$$+ \ 7.80$$

Step 2 Add the decimals.
$$47.68$$
$$+ \ 7.80$$
$$\overline{55.48}$$

The sum of 47.68 and 7.80 is 55.48.

Example 2 Find the difference of 42.17 and 15.85.

Step 1 Line up the decimal places when you write the number.
$$42.17$$
$$-15.85$$

Step 2 Subtract the decimals.
$$42.17$$
$$-15.85$$
$$\overline{26.32}$$

The difference of 42.17 and 15.85 is 26.32.

Practice Problem Find the sum of 1.245 and 3.842.

Multiply Decimals To multiply decimals, multiply the numbers like any other number, ignoring the decimal point. Count the decimal places in each factor. The product will have the same number of decimal places as the sum of the decimal places in the factors.

Example Multiply 2.4 by 5.9.

Step 1 Multiply the factors like two whole numbers.
$24 \times 59 = 1416$

Step 2 Find the sum of the number of decimal places in the factors. Each factor has one decimal place, for a sum of two decimal places.

Step 3 The product will have two decimal places.
14.16

The product of 2.4 and 5.9 is 14.16.

Practice Problem Multiply 4.6 by 2.2.

Divide Decimals When dividing decimals, change the divisor to a whole number. To do this, multiply both the divisor and the dividend by the same power of ten. Then place the decimal point in the quotient directly above the decimal point in the dividend. Then divide as you do with whole numbers.

Example Divide 8.84 by 3.4.

Step 1 Multiply both factors by 10.
$3.4 \times 10 = 34, 8.84 \times 10 = 88.4$

Step 2 Divide 88.4 by 34.

$$
\begin{array}{r}
2.6 \\
34\overline{)88.4} \\
-68 \\
\hline
204 \\
-204 \\
\hline
0
\end{array}
$$

8.84 divided by 3.4 is 2.6.

Practice Problem Divide 75.6 by 3.6.

Use Proportions

An equation that shows that two ratios are equivalent is a proportion. The ratios $\frac{2}{4}$ and $\frac{5}{10}$ are equivalent, so they can be written as $\frac{2}{4} = \frac{5}{10}$. This equation is a proportion.

When two ratios form a proportion, the cross products are equal. To find the cross products in the proportion $\frac{2}{4} = \frac{5}{10}$, multiply the 2 and the 10, and the 4 and the 5. Therefore $2 \times 10 = 4 \times 5$, or $20 = 20$.

Because you know that both proportions are equal, you can use cross products to find a missing term in a proportion. This is known as solving the proportion.

Example The heights of a tree and a pole are proportional to the lengths of their shadows. The tree casts a shadow of 24 m when a 6-m pole casts a shadow of 4 m. What is the height of the tree?

Step 1 Write a proportion.
$$\frac{\text{height of tree}}{\text{height of pole}} = \frac{\text{length of tree's shadow}}{\text{length of pole's shadow}}$$

Step 2 Substitute the known values into the proportion. Let h represent the unknown value, the height of the tree.
$$\frac{h}{6} = \frac{24}{4}$$

Step 3 Find the cross products.
$h \times 4 = 6 \times 24$

Step 4 Simplify the equation.
$4h = 144$

Step 5 Divide each side by 4.
$$\frac{4h}{4} = \frac{144}{4}$$
$$h = 36$$

The height of the tree is 36 m.

Practice Problem The ratios of the weights of two objects on the Moon and on Earth are in proportion. A rock weighing 3 N on the Moon weighs 18 N on Earth. How much would a rock that weighs 5 N on the Moon weigh on Earth?

Use Percentages

The word *percent* means "out of one hundred." It is a ratio that compares a number to 100. Suppose you read that 77 percent of the Earth's surface is covered by water. That is the same as reading that the fraction of the Earth's surface covered by water is $\frac{77}{100}$. To express a fraction as a percent, first find the equivalent decimal for the fraction. Then, multiply the decimal by 100 and add the percent symbol.

Example Express $\frac{13}{20}$ as a percent.

Step 1 Find the equivalent decimal for the fraction.

$$
\begin{array}{r}
0.65 \\
20\overline{)13.00} \\
\underline{12\ 0} \\
1\ 00 \\
\underline{1\ 00} \\
0
\end{array}
$$

Step 2 Rewrite the fraction $\frac{13}{20}$ as 0.65.

Step 3 Multiply 0.65 by 100 and add the % sign.

$0.65 \times 100 = 65 = 65\%$

So, $\frac{13}{20} = 65\%$.

This also can be solved as a proportion.

Example Express $\frac{13}{20}$ as a percent.

Step 1 Write a proportion.

$\frac{13}{20} = \frac{x}{100}$

Step 2 Find the cross products.

$1300 = 20x$

Step 3 Divide each side by 20.

$\frac{1300}{20} = \frac{20x}{20}$

$65\% = x$

Practice Problem In one year, 73 of 365 days were rainy in one city. What percent of the days in that city were rainy?

Solve One-Step Equations

A statement that two things are equal is an equation. For example, $A = B$ is an equation that states that A is equal to B.

An equation is solved when a variable is replaced with a value that makes both sides of the equation equal. To make both sides equal the inverse operation is used. Addition and subtraction are inverses, and multiplication and division are inverses.

Example 1 Solve the equation $x - 10 = 35$.

Step 1 Find the solution by adding 10 to each side of the equation.

$x - 10 = 35$

$x - 10 + 10 = 35 + 10$

$x = 45$

Step 2 Check the solution.

$x - 10 = 35$

$45 - 10 = 35$

$35 = 35$

Both sides of the equation are equal, so $x = 45$.

Example 2 In the formula $a = bc$, find the value of c if $a = 20$ and $b = 2$.

Step 1 Rearrange the formula so the unknown value is by itself on one side of the equation by dividing both sides by b.

$a = bc$

$\frac{a}{b} = \frac{bc}{b}$

$\frac{a}{b} = c$

Step 2 Replace the variables a and b with the values that are given.

$\frac{a}{b} = c$

$\frac{20}{2} = c$

$10 = c$

Step 3 Check the solution.

$a = bc$

$20 = 2 \times 10$

$20 = 20$

Both sides of the equation are equal, so $c = 10$ is the solution when $a = 20$ and $b = 2$.

Practice Problem In the formula $h = gd$, find the value of d if $g = 12.3$ and $h = 17.4$.

Math Skill Handbook

Use Statistics

The branch of mathematics that deals with collecting, analyzing, and presenting data is statistics. In statistics, there are three common ways to summarize data with a single number—the mean, the median, and the mode.

The **mean** of a set of data is the arithmetic average. It is found by adding the numbers in the data set and dividing by the number of items in the set.

The **median** is the middle number in a set of data when the data are arranged in numerical order. If there were an even number of data points, the median would be the mean of the two middle numbers.

The **mode** of a set of data is the number or item that appears most often.

Another number that often is used to describe a set of data is the range. The **range** is the difference between the largest number and the smallest number in a set of data.

A **frequency table** shows how many times each piece of data occurs, usually in a survey. **Table 2** below shows the results of a student survey on favorite color.

Table 2 Student Color Choice		
Color	**Tally**	**Frequency**
red	‖‖	4
blue	‖‖‖	5
black	‖	2
green	‖‖	3
purple	‖‖‖ ‖	7
yellow	‖‖‖ ‖	6

Based on the frequency table data, which color is the favorite?

Example The speeds (in m/s) for a race car during five different time trials are 39, 37, 44, 36, and 44.

To find the mean:

Step 1 Find the sum of the numbers.
$$39 + 37 + 44 + 36 + 44 = 200$$

Step 2 Divide the sum by the number of items, which is 5.
$$200 \div 5 = 40$$

The mean is 40 m/s.

To find the median:

Step 1 Arrange the measures from least to greatest.
36, 37, 39, 44, 44

Step 2 Determine the middle measure.
36, 37, <u>39</u>, 44, 44

The median is 39 m/s.

To find the mode:

Step 1 Group the numbers that are the same together.
44, 44, 36, 37, 39

Step 2 Determine the number that occurs most in the set.
<u>44, 44</u>, 36, 37, 39

The mode is 44 m/s.

To find the range:

Step 1 Arrange the measures from largest to smallest.
44, 44, 39, 37, 36

Step 2 Determine the largest and smallest measures in the set.
<u>44</u>, 44, 39, 37, <u>36</u>

Step 3 Find the difference between the largest and smallest measures.
$$44 - 36 = 8$$

The range is 8 m/s.

Practice Problem Find the mean, median, mode, and range for the data set 8, 4, 12, 8, 11, 14, 16.

Use Geometry

The branch of mathematics that deals with the measurement, properties, and relationships of points, lines, angles, surfaces, and solids is called geometry.

Perimeter The **perimeter** (P) is the distance around a geometric figure. To find the perimeter of a rectangle, add the length and width and multiply that sum by two, or $2(l + w)$. To find perimeters of irregular figures, add the length of the sides.

Example 1 Find the perimeter of a rectangle that is 3 m long and 5 m wide.

Step 1 You know that the perimeter is 2 times the sum of the width and length.
$P = 2(3\text{ m} + 5\text{ m})$

Step 2 Find the sum of the width and length.
$P = 2(8\text{ m})$

Step 3 Multiply by 2.
$P = 16\text{ m}$

The perimeter is 16 m.

Example 2 Find the perimeter of a shape with sides measuring 2 cm, 5 cm, 6 cm, 3 cm.

Step 1 You know that the perimeter is the sum of all the sides.
$P = 2 + 5 + 6 + 3$

Step 2 Find the sum of the sides.
$P = 2 + 5 + 6 + 3$
$P = 16$

The perimeter is 16 cm.

Practice Problem Find the perimeter of a rectangle with a length of 18 m and a width of 7 m.

Practice Problem Find the perimeter of a triangle measuring 1.6 cm by 2.4 cm by 2.4 cm.

Area of a Rectangle The **area** (A) is the number of square units needed to cover a surface. To find the area of a rectangle, multiply the length times the width, or $l \times w$. When finding area, the units also are multiplied. Area is given in square units.

Example Find the area of a rectangle with a length of 1 cm and a width of 10 cm.

Step 1 You know that the area is the length multiplied by the width.
$A = (1\text{ cm} \times 10\text{ cm})$

Step 2 Multiply the length by the width. Also multiply the units.
$A = 10\text{ cm}^2$

The area is 10 cm^2.

Practice Problem Find the area of a square whose sides measure 4 m.

Area of a Triangle To find the area of a triangle, use the formula:

$$A = \frac{1}{2}(\text{base} \times \text{height})$$

The base of a triangle can be any of its sides. The height is the perpendicular distance from a base to the opposite endpoint, or vertex.

Example Find the area of a triangle with a base of 18 m and a height of 7 m.

Step 1 You know that the area is $\frac{1}{2}$ the base times the height.
$A = \frac{1}{2}(18\text{ m} \times 7\text{ m})$

Step 2 Multiply $\frac{1}{2}$ by the product of 18×7. Multiply the units.
$A = \frac{1}{2}(126\text{ m}^2)$
$A = 63\text{ m}^2$

The area is 63 m^2.

Practice Problem Find the area of a triangle with a base of 27 cm and a height of 17 cm.

Circumference of a Circle The **diameter** (d) of a circle is the distance across the circle through its center, and the **radius** (r) is the distance from the center to any point on the circle. The radius is half of the diameter. The distance around the circle is called the **circumference** (C). The formula for finding the circumference is:

$$C = 2\pi r \ \ or \ \ C = \pi d$$

The circumference divided by the diameter is always equal to 3.1415926... This nonterminating and nonrepeating number is represented by the Greek letter π (pi). An approximation often used for π is 3.14.

Example 1 Find the circumference of a circle with a radius of 3 m.

Step 1 You know the formula for the circumference is 2 times the radius times π.
$$C = 2\pi(3)$$

Step 2 Multiply 2 times the radius.
$$C = 6\pi$$

Step 3 Multiply by π.
$$C = 19 \text{ m}$$

The circumference is 19 m.

Example 2 Find the circumference of a circle with a diameter of 24.0 cm.

Step 1 You know the formula for the circumference is the diameter times π.
$$C = \pi(24.0)$$

Step 2 Multiply the diameter by π.
$$C = 75.4 \text{ cm}$$

The circumference is 75.4 cm.

Practice Problem Find the circumference of a circle with a radius of 19 cm.

Area of a Circle The formula for the area of a circle is:
$$A = \pi r^2$$

Example 1 Find the area of a circle with a radius of 4.0 cm.

Step 1 $A = \pi(4.0)^2$

Step 2 Find the square of the radius.
$$A = 16\pi$$

Step 3 Multiply the square of the radius by π.
$$A = 50 \text{ cm}^2$$

The area of the circle is 50 cm^2.

Example 2 Find the area of a circle with a radius of 225 m.

Step 1 $A = \pi(225)^2$

Step 2 Find the square of the radius.
$$A = 50625\pi$$

Step 3 Multiply the square of the radius by π.
$$A = 158962.5$$

The area of the circle is 158,962 m^2.

Example 3 Find the area of a circle whose diameter is 20.0 mm.

Step 1 You know the formula for the area of a circle is the square of the radius times π, and that the radius is half of the diameter.
$$A = \pi\left(\frac{20.0}{2}\right)^2$$

Step 2 Find the radius.
$$A = \pi(10.0)^2$$

Step 3 Find the square of the radius.
$$A = 100\pi$$

Step 4 Multiply the square of the radius by π.
$$A = 314 \text{ mm}^2$$

The area is 314 mm^2.

Practice Problem Find the area of a circle with a radius of 16 m.

Math Skill Handbook

Volume The measure of space occupied by a solid is the **volume** (V). To find the volume of a rectangular solid multiply the length times width times height, or $V = l \times w \times h$. It is measured in cubic units, such as cubic centimeters (cm^3).

Example Find the volume of a rectangular solid with a length of 2.0 m, a width of 4.0 m, and a height of 3.0 m.

Step 1 You know the formula for volume is the length times the width times the height.
$$V = 2.0 \text{ m} \times 4.0 \text{ m} \times 3.0 \text{ m}$$

Step 2 Multiply the length times the width times the height.
$$V = 24 \text{ m}^3$$

The volume is 24 m³.

Practice Problem Find the volume of a rectangular solid that is 8 m long, 4 m wide, and 4 m high.

To find the volume of other solids, multiply the area of the base times the height.

Example 1 Find the volume of a solid that has a triangular base with a length of 8.0 m and a height of 7.0 m. The height of the entire solid is 15.0 m.

Step 1 You know that the base is a triangle, and the area of a triangle is $\frac{1}{2}$ the base times the height, and the volume is the area of the base times the height.
$$V = \left[\frac{1}{2}(b \times h)\right] \times 15$$

Step 2 Find the area of the base.
$$V = \left[\frac{1}{2}(8 \times 7)\right] \times 15$$
$$V = \left(\frac{1}{2} \times 56\right) \times 15$$

Step 3 Multiply the area of the base by the height of the solid.
$$V = 28 \times 15$$
$$V = 420 \text{ m}^3$$

The volume is 420 m³.

Example 2 Find the volume of a cylinder that has a base with a radius of 12.0 cm, and a height of 21.0 cm.

Step 1 You know that the base is a circle, and the area of a circle is the square of the radius times π, and the volume is the area of the base times the height.
$$V = (\pi r^2) \times 21$$
$$V = (\pi 12^2) \times 21$$

Step 2 Find the area of the base.
$$V = 144\pi \times 21$$
$$V = 452 \times 21$$

Step 3 Multiply the area of the base by the height of the solid.
$$V = 9490 \text{ cm}^3$$

The volume is 9490 cm³.

Example 3 Find the volume of a cylinder that has a diameter of 15 mm and a height of 4.8 mm.

Step 1 You know that the base is a circle with an area equal to the square of the radius times π. The radius is one-half the diameter. The volume is the area of the base times the height.
$$V = (\pi r^2) \times 4.8$$
$$V = \left[\pi\left(\frac{1}{2} \times 15\right)^2\right] \times 4.8$$
$$V = (\pi 7.5^2) \times 4.8$$

Step 2 Find the area of the base.
$$V = 56.25\pi \times 4.8$$
$$V = 176.63 \times 4.8$$

Step 3 Multiply the area of the base by the height of the solid.
$$V = 847.8$$

The volume is 847.8 mm³.

Practice Problem Find the volume of a cylinder with a diameter of 7 cm in the base and a height of 16 cm.

Science Applications

Measure in SI

The metric system of measurement was developed in 1795. A modern form of the metric system, called the International System (SI), was adopted in 1960 and provides the standard measurements that all scientists around the world can understand.

The SI system is convenient because unit sizes vary by powers of 10. Prefixes are used to name units. Look at **Table 3** for some common SI prefixes and their meanings.

Table 3 Some SI Prefixes			
Prefix	**Symbol**	**Meaning**	
kilo-	k	1,000	thousand
hecto-	h	100	hundred
deka-	da	10	ten
deci-	d	0.1	tenth
centi-	c	0.01	hundredth
milli-	m	0.001	thousandth

Example How many grams equal one kilogram?

Step 1 Find the prefix *kilo* in **Table 3.**

Step 2 Using **Table 3,** determine the meaning of *kilo.* According to the table, it means 1,000. When the prefix *kilo* is added to a unit, it means that there are 1,000 of the units in a "*kilo*unit."

Step 3 Apply the prefix to the units in the question. The units in the question are grams. There are 1,000 grams in a kilogram.

Practice Problem Is a milligram larger or smaller than a gram? How many of the smaller units equal one larger unit? What fraction of the larger unit does one smaller unit represent?

Dimensional Analysis

Convert SI Units In science, quantities such as length, mass, and time sometimes are measured using different units. A process called dimensional analysis can be used to change one unit of measure to another. This process involves multiplying your starting quantity and units by one or more conversion factors. A conversion factor is a ratio equal to one and can be made from any two equal quantities with different units. If 1,000 mL equal 1 L then two ratios can be made.

$$\frac{1,000 \text{ mL}}{1 \text{ L}} = \frac{1 \text{ L}}{1,000 \text{ mL}} = 1$$

One can covert between units in the SI system by using the equivalents in **Table 3** to make conversion factors.

Example 1 How many cm are in 4 m?

Step 1 Write conversion factors for the units given. From **Table 3,** you know that 100 cm = 1 m. The conversion factors are

$$\frac{100 \text{ cm}}{1 \text{ m}} \quad and \quad \frac{1 \text{ m}}{100 \text{ cm}}$$

Step 2 Decide which conversion factor to use. Select the factor that has the units you are converting from (m) in the denominator and the units you are converting to (cm) in the numerator.

$$\frac{100 \text{ cm}}{1 \text{ m}}$$

Step 3 Multiply the starting quantity and units by the conversion factor. Cancel the starting units with the units in the denominator. There are 400 cm in 4 m.

$$4 \text{ m} \times \frac{100 \text{ cm}}{1 \text{ m}} = 400 \text{ cm}$$

Practice Problem How many milligrams are in one kilogram? (Hint: You will need to use two conversion factors from **Table 3.**)

Table 4 Unit System Equivalents

Type of Measurement	Equivalent
Length	1 in = 2.54 cm
	1 yd = 0.91 m
	1 mi = 1.61 km
Mass and Weight*	1 oz = 28.35 g
	1 lb = 0.45 kg
	1 ton (short) = 0.91 tonnes (metric tons)
	1 lb = 4.45 N
Volume	$1 \text{ in}^3 = 16.39 \text{ cm}^3$
	1 qt = 0.95 L
	1 gal = 3.78 L
Area	$1 \text{ in}^2 = 6.45 \text{ cm}^2$
	$1 \text{ yd}^2 = 0.83 \text{ m}^2$
	$1 \text{ mi}^2 = 2.59 \text{ km}^2$
	1 acre = 0.40 hectares
Temperature	$°C = \dfrac{(°F - 32)}{1.8}$
	$K = °C + 273$

*Weight is measured in standard Earth gravity.

Convert Between Unit Systems Table 4 gives a list of equivalents that can be used to convert between English and SI units.

Example If a meterstick has a length of 100 cm, how long is the meterstick in inches?

Step 1 Write the conversion factors for the units given. From **Table 4,** 1 in = 2.54 cm.

$$\frac{1 \text{ in}}{2.54 \text{ cm}} \quad and \quad \frac{2.54 \text{ cm}}{1 \text{ in}}$$

Step 2 Determine which conversion factor to use. You are converting from cm to in. Use the conversion factor with cm on the bottom.

$$\frac{1 \text{ in}}{2.54 \text{ cm}}$$

Step 3 Multiply the starting quantity and units by the conversion factor. Cancel the starting units with the units in the denominator. Round your answer based on the number of significant figures in the conversion factor.

$$100 \cancel{\text{ cm}} \times \frac{1 \text{ in}}{2.54 \cancel{\text{ cm}}} = 39.37 \text{ in}$$

The meterstick is 39.4 in long.

Practice Problem A book has a mass of 5 lbs. What is the mass of the book in kg?

Practice Problem Use the equivalent for in and cm (1 in = 2.54 cm) to show how $1 \text{ in}^3 = 16.39 \text{ cm}^3$.

Precision and Significant Digits

When you make a measurement, the value you record depends on the precision of the measuring instrument. This precision is represented by the number of significant digits recorded in the measurement. When counting the number of significant digits, all digits are counted except zeros at the end of a number with no decimal point such as 2,050, and zeros at the beginning of a decimal such as 0.03020. When adding or subtracting numbers with different precision, round the answer to the smallest number of decimal places of any number in the sum or difference. When multiplying or dividing, the answer is rounded to the smallest number of significant digits of any number being multiplied or divided.

Example The lengths 5.28 and 5.2 are measured in meters. Find the sum of these lengths and record your answer using the correct number of significant digits.

Step 1 Find the sum.

5.28 m	2 digits after the decimal
+ 5.2 m	1 digit after the decimal
10.48 m	

Step 2 Round to one digit after the decimal because the least number of digits after the decimal of the numbers being added is 1.

The sum is 10.5 m.

Practice Problem How many significant digits are in the measurement 7,071,301 m? How many significant digits are in the measurement 0.003010 g?

Practice Problem Multiply 5.28 and 5.2 using the rule for multiplying and dividing. Record the answer using the correct number of significant digits.

Scientific Notation

Many times numbers used in science are very small or very large. Because these numbers are difficult to work with scientists use scientific notation. To write numbers in scientific notation, move the decimal point until only one non-zero digit remains on the left. Then count the number of places you moved the decimal point and use that number as a power of ten. For example, the average distance from the Sun to Mars is 227,800,000,000 m. In scientific notation, this distance is 2.278×10^{11} m. Because you moved the decimal point to the left, the number is a positive power of ten.

The mass of an electron is about 0.000 000 000 000 000 000 000 000 000 000 911 kg. Expressed in scientific notation, this mass is 9.11×10^{-31} kg. Because the decimal point was moved to the right, the number is a negative power of ten.

Example Earth is 149,600,000 km from the Sun. Express this in scientific notation.

Step 1 Move the decimal point until one non-zero digit remains on the left.
1.496 000 00

Step 2 Count the number of decimal places you have moved. In this case, eight.

Step 3 Show that number as a power of ten, 10^8.

The Earth is 1.496×10^8 km from the Sun.

Practice Problem How many significant digits are in 149,600,000 km? How many significant digits are in 1.496×10^8 km?

Practice Problem Parts used in a high performance car must be measured to 7×10^{-6} m. Express this number as a decimal.

Practice Problem A CD is spinning at 539 revolutions per minute. Express this number in scientific notation.

Make and Use Graphs

Data in tables can be displayed in a graph—a visual representation of data. Common graph types include line graphs, bar graphs, and circle graphs.

Line Graph A line graph shows a relationship between two variables that change continuously. The independent variable is changed and is plotted on the *x*-axis. The dependent variable is observed, and is plotted on the *y*-axis.

Example Draw a line graph of the data below from a cyclist in a long-distance race.

Table 5 Bicycle Race Data	
Time (h)	**Distance (km)**
0	0
1	8
2	16
3	24
4	32
5	40

Step 1 Determine the *x*-axis and *y*-axis variables. Time varies independently of distance and is plotted on the *x*-axis. Distance is dependent on time and is plotted on the *y*-axis.

Step 2 Determine the scale of each axis. The *x*-axis data ranges from 0 to 5. The *y*-axis data ranges from 0 to 40.

Step 3 Using graph paper, draw and label the axes. Include units in the labels.

Step 4 Draw a point at the intersection of the time value on the *x*-axis and corresponding distance value on the *y*-axis. Connect the points and label the graph with a title, as shown in **Figure 20.**

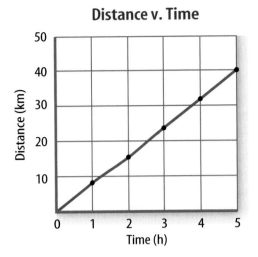

Distance v. Time

Figure 20 This line graph shows the relationship between distance and time during a bicycle ride.

Practice Problem A puppy's shoulder height is measured during the first year of her life. The following measurements were collected: (3 mo, 52 cm), (6 mo, 72 cm), (9 mo, 83 cm), (12 mo, 86 cm). Graph this data.

Find a Slope The slope of a straight line is the ratio of the vertical change, rise, to the horizontal change, run.

$$\text{Slope} = \frac{\text{vertical change (rise)}}{\text{horizontal change (run)}} = \frac{\text{change in } y}{\text{change in } x}$$

Example Find the slope of the graph in **Figure 20.**

Step 1 You know that the slope is the change in *y* divided by the change in *x*.
$$\text{Slope} = \frac{\text{change in } y}{\text{change in } x}$$

Step 2 Determine the data points you will be using. For a straight line, choose the two sets of points that are the farthest apart.
$$\text{Slope} = \frac{(40-0) \text{ km}}{(5-0) \text{ hr}}$$

Step 3 Find the change in *y* and *x*.
$$\text{Slope} = \frac{40 \text{ km}}{5 \text{h}}$$

Step 4 Divide the change in *y* by the change in *x*.
$$\text{Slope} = \frac{8 \text{ km}}{\text{h}}$$

The slope of the graph is 8 km/h.

Bar Graph To compare data that does not change continuously you might choose a bar graph. A bar graph uses bars to show the relationships between variables. The *x*-axis variable is divided into parts. The parts can be numbers such as years, or a category such as a type of animal. The *y*-axis is a number and increases continuously along the axis.

Example A recycling center collects 4.0 kg of aluminum on Monday, 1.0 kg on Wednesday, and 2.0 kg on Friday. Create a bar graph of this data.

Step 1 Select the *x*-axis and *y*-axis variables. The measured numbers (the masses of aluminum) should be placed on the *y*-axis. The variable divided into parts (collection days) is placed on the *x*-axis.

Step 2 Create a graph grid like you would for a line graph. Include labels and units.

Step 3 For each measured number, draw a vertical bar above the *x*-axis value up to the *y*-axis value. For the first data point, draw a vertical bar above Monday up to 4.0 kg.

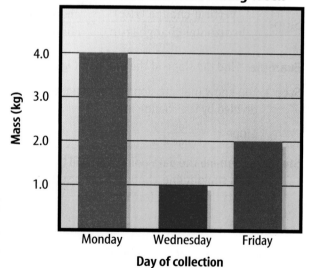

Aluminum Collected During Week

Practice Problem Draw a bar graph of the gases in air: 78% nitrogen, 21% oxygen, 1% other gases.

Circle Graph To display data as parts of a whole, you might use a circle graph. A circle graph is a circle divided into sections that represent the relative size of each piece of data. The entire circle represents 100%, half represents 50%, and so on.

Example Air is made up of 78% nitrogen, 21% oxygen, and 1% other gases. Display the composition of air in a circle graph.

Step 1 Multiply each percent by 360° and divide by 100 to find the angle of each section in the circle.

$$78\% \times \frac{360°}{100} = 280.8°$$

$$21\% \times \frac{360°}{100} = 75.6°$$

$$1\% \times \frac{360°}{100} = 3.6°$$

Step 2 Use a compass to draw a circle and to mark the center of the circle. Draw a straight line from the center to the edge of the circle.

Step 3 Use a protractor and the angles you calculated to divide the circle into parts. Place the center of the protractor over the center of the circle and line the base of the protractor over the straight line.

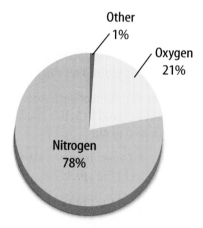

Practice Problem Draw a circle graph to represent the amount of aluminum collected during the week shown in the bar graph to the left.

Weather Map Symbols

Sample Station Model

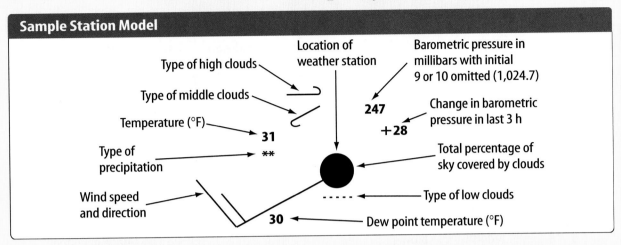

Precipitation	Wind Speed and Direction	Sky Coverage	Some Types of High Clouds
☰ Fog	○ 0 calm	○ No cover	Scattered cirrus
★ Snow	1–2 knots	◐ 1/10 or less	Dense cirrus in patches
● Rain	3–7 knots	◕ 2/10 to 3/10	Veil of cirrus covering entire sky
Thunderstorm	8–12 knots	◑ 4/10	Cirrus not covering entire sky
, Drizzle	13–17 knots	◑ –	
▽ Showers	18–22 knots	◕ 6/10	
	23–27 knots	◕ 7/10	
	48–52 knots	◕ Overcast with openings	
	1 knot = 1.852 km/h	● Completely overcast	

Sample Plotted Report at Each Station

Some Types of Middle Clouds	Some Types of Low Clouds	Fronts and Pressure Systems	
Thin altostratus layer	⌒ Cumulus of fair weather	(H) or High (L) or Low	Center of high- or low-pressure system
Thick altostratus layer	⌄ Stratocumulus	▲▲▲▲	Cold front
Thin altostratus in patches	- - - - - Fractocumulus of bad weather	●●●●	Warm front
Thin altostratus in bands	—— Stratus of fair weather	▲●▲●	Occluded front
		●▲●▽	Stationary front

Minerals

Mineral (formula)	Color	Streak	Hardness	Breakage Pattern	Uses and Other Properties
Graphite (C)	black to gray	black to gray	1–1.5	basal cleavage (scales)	pencil lead, lubricants for locks, rods to control some small nuclear reactions, battery poles
Galena (PbS)	gray	gray to black	2.5	cubic cleavage perfect	source of lead, used for pipes, shields for X rays, fishing equipment sinkers
Hematite (Fe_2O_3)	black or reddish-brown	reddish-brown	5.5–6.5	irregular fracture	source of iron; converted to pig iron, made into steel
Magnetite (Fe_3O_4)	black	black	6	conchoidal fracture	source of iron, attracts a magnet
Pyrite (FeS_2)	light, brassy, yellow	greenish-black	6–6.5	uneven fracture	fool's gold
Talc ($Mg_3 Si_4 O_{10} (OH)_2$)	white, greenish	white	1	cleavage in one direction	used for talcum powder, sculptures, paper, and tabletops
Gypsum ($CaSO_4 \cdot 2H_2O$)	colorless, gray, white, brown	white	2	basal cleavage	used in plaster of paris and dry wall for building construction
Sphalerite (ZnS)	brown, reddish-brown, greenish	light to dark brown	3.5–4	cleavage in six directions	main ore of zinc; used in paints, dyes, and medicine
Muscovite ($KAl_3Si_3 O_{10}(OH)_2$)	white, light gray, yellow, rose, green	colorless	2–2.5	basal cleavage	occurs in large, flexible plates; used as an insulator in electrical equipment, lubricant
Biotite ($K(Mg,Fe)_3 (AlSi_3O_{10}) (OH)_2$)	black to dark brown	colorless	2.5–3	basal cleavage	occurs in large, flexible plates
Halite (NaCl)	colorless, red, white, blue	colorless	2.5	cubic cleavage	salt; soluble in water; a preservative

Minerals

Minerals					
Mineral (formula)	Color	Streak	Hardness	Breakage Pattern	Uses and Other Properties
Calcite $(CaCO_3)$	colorless, white, pale blue	colorless, white	3	cleavage in three directions	fizzes when HCl is added; used in cements and other building materials
Dolomite $(CaMg (CO_3)_2)$	colorless, white, pink, green, gray, black	white	3.5–4	cleavage in three directions	concrete and cement; used as an ornamental building stone
Fluorite (CaF_2)	colorless, white, blue, green, red, yellow, purple	colorless	4	cleavage in four directions	used in the manufacture of optical equipment; glows under ultraviolet light
Hornblende $(CaNa)_{2-3}$ $(Mg,Al,$ $Fe)_5-(Al,Si)_2$ Si_6O_{22} $(OH)_2)$	green to black	gray to white	5–6	cleavage in two directions	will transmit light on thin edges; 6-sided cross section
Feldspar $(KAlSi_3O_8)$ $(NaAl$ $Si_3O_8),$ $(CaAl_2Si_2$ $O_8)$	colorless, white to gray, green	colorless	6	two cleavage planes meet at 90° angle	used in the manufacture of ceramics
Augite $(((Ca,Na)$ (Mg,Fe,Al) $(Al,Si)_2 O_6)$	black	colorless	6	cleavage in two directions	square or 8-sided cross section
Olivine $((Mg,Fe)_2$ $SiO_4)$	olive, green	none	6.5–7	conchoidal fracture	gemstones, refractory sand
Quartz (SiO_2)	colorless, various colors	none	7	conchoidal fracture	used in glass manufacture, electronic equipment, radios, computers, watches, gemstones

Rocks

Rocks		
Rock Type	**Rock Name**	**Characteristics**
Igneous (intrusive)	Granite	Large mineral grains of quartz, feldspar, hornblende, and mica. Usually light in color.
	Diorite	Large mineral grains of feldspar, hornblende, and mica. Less quartz than granite. Intermediate in color.
	Gabbro	Large mineral grains of feldspar, augite, and olivine. No quartz. Dark in color.
Igneous (extrusive)	Rhyolite	Small mineral grains of quartz, feldspar, hornblende, and mica, or no visible grains. Light in color.
	Andesite	Small mineral grains of feldspar, hornblende, and mica or no visible grains. Intermediate in color.
	Basalt	Small mineral grains of feldspar, augite, and possibly olivine or no visible grains. No quartz. Dark in color.
	Obsidian	Glassy texture. No visible grains. Volcanic glass. Fracture looks like broken glass.
	Pumice	Frothy texture. Floats in water. Usually light in color.
Sedimentary (detrital)	Conglomerate	Coarse grained. Gravel or pebble-size grains.
	Sandstone	Sand-sized grains 1/16 to 2 mm.
	Siltstone	Grains are smaller than sand but larger than clay.
	Shale	Smallest grains. Often dark in color. Usually platy.
Sedimentary (chemical or organic)	Limestone	Major mineral is calcite. Usually forms in oceans and lakes. Often contains fossils.
	Coal	Forms in swampy areas. Compacted layers of organic material, mainly plant remains.
Sedimentary (chemical)	Rock Salt	Commonly forms by the evaporation of seawater.
Metamorphic (foliated)	Gneiss	Banding due to alternate layers of different minerals, of different colors. Parent rock often is granite.
	Schist	Parallel arrangement of sheetlike minerals, mainly micas. Forms from different parent rocks.
	Phyllite	Shiny or silky appearance. May look wrinkled. Common parent rocks are shale and slate.
	Slate	Harder, denser, and shinier than shale. Common parent rock is shale.
Metamorphic (nonfoliated)	Marble	Calcite or dolomite. Common parent rock is limestone.
	Soapstone	Mainly of talc. Soft with greasy feel.
	Quartzite	Hard with interlocking quartz crystals. Common parent rock is sandstone.

Topographic Map Symbols

Topographic Map Symbols

▬▬▬▬	Primary highway, hard surface	〜〜	Index contour
▬▭▬	Secondary highway, hard surface	··········	Supplementary contour
═══	Light-duty road, hard or improved surface	〜〜	Intermediate contour
=========	Unimproved road	⬭	Depression contours
+—+—+—+	Railroad: single track		
+╫—╫—╫+	Railroad: multiple track	▬ ▬ ▬	Boundaries: national
+╪+╪+╪+	Railroads in juxtaposition	▬ — ▬	State
		— — -	County, parish, municipal
▪▫▬▦	Buildings	— — -	Civil township, precinct, town, barrio
♪▫✝ cem	Schools, church, and cemetery	— · — ·	Incorporated city, village, town, hamlet
▫▭▨▨	Buildings (barn, warehouse, etc.)	· — · — ·	Reservation, national or state
o o	Wells other than water (labeled as to type)	---------	Small park, cemetery, airport, etc.
●●●⦸	Tanks: oil, water, etc. (labeled only if water)	— ·· — ··	Land grant
⊙ ⚥	Located or landmark object; windmill	▬▬▬	Township or range line, U.S. land survey
⤬ ✕	Open pit, mine, or quarry; prospect	— — —	Township or range line, approximate location
▦	Marsh (swamp)		
▦	Wooded marsh	～～	Perennial streams
▫	Woods or brushwood	→ ←	Elevated aqueduct
▫	Vineyard	o ○〜	Water well and spring
▫	Land subject to controlled inundation	〜⊬〜	Small rapids
▫	Submerged marsh	▨	Large rapids
▦	Mangrove	▧	Intermittent lake
▫	Orchard	〜〜	Intermittent stream
▫	Scrub	→==←	Aqueduct tunnel
▦	Urban area	▨	Glacier
		〜⊬	Small falls
x7369	Spot elevation	▨	Large falls
670	Water elevation	▨	Dry lake bed

PERIODIC TABLE OF THE ELEMENTS

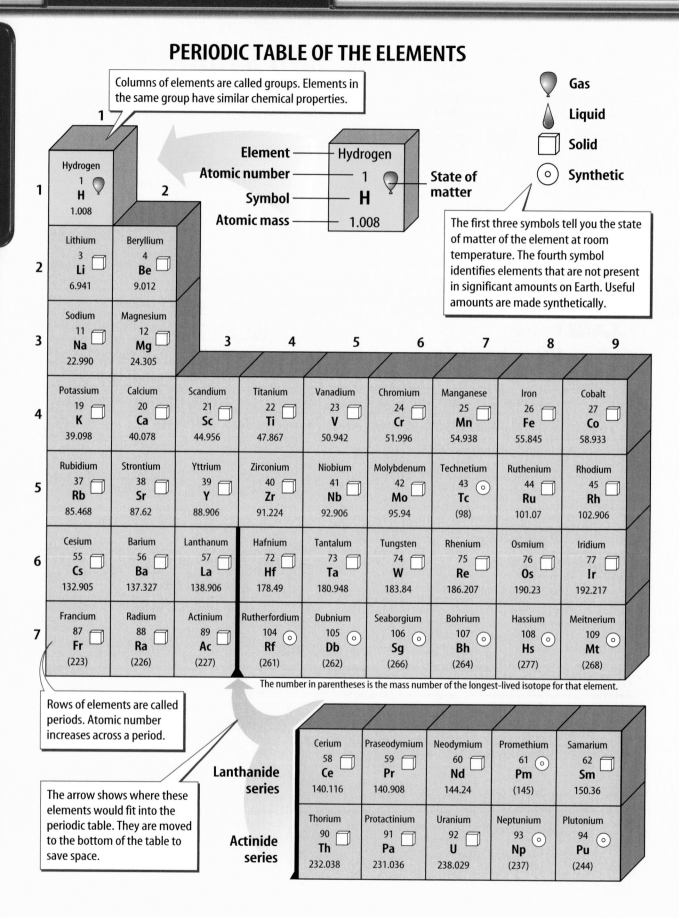

Columns of elements are called groups. Elements in the same group have similar chemical properties.

Element ——— Hydrogen
Atomic number ——— 1
Symbol ——— H
Atomic mass ——— 1.008

State of matter

The first three symbols tell you the state of matter of the element at room temperature. The fourth symbol identifies elements that are not present in significant amounts on Earth. Useful amounts are made synthetically.

Gas
Liquid
Solid
Synthetic

1

1		
Hydrogen		
1		
H		
1.008		

2

Lithium	Beryllium
3	4
Li	**Be**
6.941	9.012

3

Sodium	Magnesium
11	12
Na	**Mg**
22.990	24.305

	3	4	5	6	7	8	9

4

Potassium	Calcium	Scandium	Titanium	Vanadium	Chromium	Manganese	Iron	Cobalt
19	20	21	22	23	24	25	26	27
K	**Ca**	**Sc**	**Ti**	**V**	**Cr**	**Mn**	**Fe**	**Co**
39.098	40.078	44.956	47.867	50.942	51.996	54.938	55.845	58.933

5

Rubidium	Strontium	Yttrium	Zirconium	Niobium	Molybdenum	Technetium	Ruthenium	Rhodium
37	38	39	40	41	42	43	44	45
Rb	**Sr**	**Y**	**Zr**	**Nb**	**Mo**	**Tc**	**Ru**	**Rh**
85.468	87.62	88.906	91.224	92.906	95.94	(98)	101.07	102.906

6

Cesium	Barium	Lanthanum	Hafnium	Tantalum	Tungsten	Rhenium	Osmium	Iridium
55	56	57	72	73	74	75	76	77
Cs	**Ba**	**La**	**Hf**	**Ta**	**W**	**Re**	**Os**	**Ir**
132.905	137.327	138.906	178.49	180.948	183.84	186.207	190.23	192.217

7

Francium	Radium	Actinium	Rutherfordium	Dubnium	Seaborgium	Bohrium	Hassium	Meitnerium
87	88	89	104	105	106	107	108	109
Fr	**Ra**	**Ac**	**Rf**	**Db**	**Sg**	**Bh**	**Hs**	**Mt**
(223)	(226)	(227)	(261)	(262)	(266)	(264)	(277)	(268)

The number in parentheses is the mass number of the longest-lived isotope for that element.

Rows of elements are called periods. Atomic number increases across a period.

The arrow shows where these elements would fit into the periodic table. They are moved to the bottom of the table to save space.

Lanthanide series

Cerium	Praseodymium	Neodymium	Promethium	Samarium
58	59	60	61	62
Ce	**Pr**	**Nd**	**Pm**	**Sm**
140.116	140.908	144.24	(145)	150.36

Actinide series

Thorium	Protactinium	Uranium	Neptunium	Plutonium
90	91	92	93	94
Th	**Pa**	**U**	**Np**	**Pu**
232.038	231.036	238.029	(237)	(244)

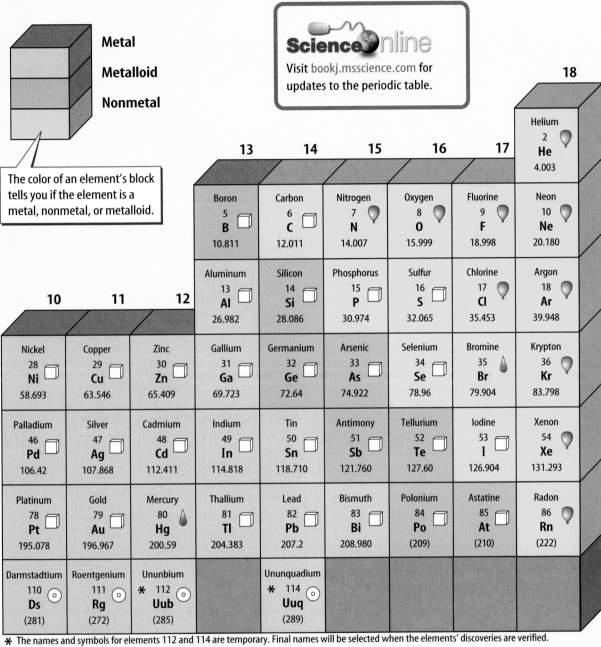

Metal

Metalloid

Nonmetal

The color of an element's block tells you if the element is a metal, nonmetal, or metalloid.

Science Online
Visit bookj.msscience.com for updates to the periodic table.

		13	14	15	16	17	18
							Helium 2 **He** 4.003
		Boron 5 **B** 10.811	Carbon 6 **C** 12.011	Nitrogen 7 **N** 14.007	Oxygen 8 **O** 15.999	Fluorine 9 **F** 18.998	Neon 10 **Ne** 20.180
10	11	12	Aluminum 13 **Al** 26.982	Silicon 14 **Si** 28.086	Phosphorus 15 **P** 30.974	Sulfur 16 **S** 32.065	Chlorine 17 **Cl** 35.453	Argon 18 **Ar** 39.948

10	11	12	13	14	15	16	17	18
Nickel 28 **Ni** 58.693	Copper 29 **Cu** 63.546	Zinc 30 **Zn** 65.409	Gallium 31 **Ga** 69.723	Germanium 32 **Ge** 72.64	Arsenic 33 **As** 74.922	Selenium 34 **Se** 78.96	Bromine 35 **Br** 79.904	Krypton 36 **Kr** 83.798
Palladium 46 **Pd** 106.42	Silver 47 **Ag** 107.868	Cadmium 48 **Cd** 112.411	Indium 49 **In** 114.818	Tin 50 **Sn** 118.710	Antimony 51 **Sb** 121.760	Tellurium 52 **Te** 127.60	Iodine 53 **I** 126.904	Xenon 54 **Xe** 131.293
Platinum 78 **Pt** 195.078	Gold 79 **Au** 196.967	Mercury 80 **Hg** 200.59	Thallium 81 **Tl** 204.383	Lead 82 **Pb** 207.2	Bismuth 83 **Bi** 208.980	Polonium 84 **Po** (209)	Astatine 85 **At** (210)	Radon 86 **Rn** (222)
Darmstadtium 110 **Ds** (281)	Roentgenium 111 **Rg** (272)	Ununbium ✳ 112 **Uub** (285)		Ununquadium ✳ 114 **Uuq** (289)				

✳ The names and symbols for elements 112 and 114 are temporary. Final names will be selected when the elements' discoveries are verified.

Europium 63 **Eu** 151.964	Gadolinium 64 **Gd** 157.25	Terbium 65 **Tb** 158.925	Dysprosium 66 **Dy** 162.500	Holmium 67 **Ho** 164.930	Erbium 68 **Er** 167.259	Thulium 69 **Tm** 168.934	Ytterbium 70 **Yb** 173.04	Lutetium 71 **Lu** 174.967
Americium 95 **Am** (243)	Curium 96 **Cm** (247)	Berkelium 97 **Bk** (247)	Californium 98 **Cf** (251)	Einsteinium 99 **Es** (252)	Fermium 100 **Fm** (257)	Mendelevium 101 **Md** (258)	Nobelium 102 **No** (259)	Lawrencium 103 **Lr** (262)

Glossary/Glosario

Cómo usar el glosario en español:
1. Busca el término en inglés que desees encontrar.
2. El término en español, junto con la definición, se encuentran en la columna de la derecha.

Pronunciation Key

Use the following key to help you sound out words in the glossary.

a . . . . back (BAK)	ew . . . . food (FEWD)
ay . . . . day (DAY)	yoo . . . . pure (PYOOR)
ah . . . . father (FAH thur)	yew . . . . few (FYEW)
ow . . . . flower (FLOW ur)	uh . . . . comma (CAH muh)
ar . . . . car (CAR)	u (+ con) . . . . . . rub (RUB)
e . . . . less (LES)	sh . . . . shelf (SHELF)
ee . . . . leaf (LEEF)	ch . . . . nature (NAY chur)
ih . . . . trip (TRIHP)	g . . . . gift (GIHFT)
i (i + con + e) . . idea (i DEE uh)	j . . . . gem (JEM)
oh . . . . go (GOH)	ing . . . . sing (SING)
aw . . . . soft (SAWFT)	zh . . . . vision (VIH zhun)
or . . . . orbit (OR buht)	k . . . . cake (KAYK)
oy . . . . coin (COYN)	s . . . . seed, cent (SEED, SENT)
oo . . . . foot (FOOT)	z . . . . zone, raise (ZOHN, RAYZ)

English — A — Español

absolute magnitude: measure of the amount of light a star actually gives off. (p. 106)

apparent magnitude: measure of the amount of light from a star that is received on Earth. (p. 106)

asteroid: a piece of rock or metal made up of material similar to that which formed the planets; mostly found in the asteroid belt between the orbits of Mars and Jupiter. (p. 92)

axis: imaginary vertical line that cuts through the center of Earth and around which Earth spins. (p. 41)

magnitud absoluta: medida de la cantidad real de luz que genera una estrella. (p. 106)

magnitud aparente: medida de la cantidad de luz recibida en la Tierra desde una estrella. (p. 106)

asteroide: pedazo de roca o metal formado de material similar al que forma los planetas; se encuentran principalmente en el cinturón de asteroides entre las órbitas de Marte y Júpiter. (p. 92)

eje: línea vertical imaginaria que atraviesa el centro de la Tierra y alrededor de la cual gira ésta. (p. 41)

B

Big Bang theory: states that about 13.7 billion years ago, the universe began with a huge, fiery explosion. (p. 125)

black hole: final stage in the evolution of a very massive star, where the core's mass collapses to a point that it's gravity is so strong that not even light can escape. (p. 118)

teoría de la Gran Explosión: establece que hace aproximadamente 13.7 billones de años el universo se originó con una enorme explosión. (p. 125)

agujero negro: etapa final en la evolución de una estrella masiva, en donde la masa del núcleo se colapsa hasta el punto de que su gravedad es tan fuerte que ni siquiera la luz puede escapar. (p. 118)

C

chromosphere: layer of the Sun's atmosphere above the photosphere. (p. 109)

cromosfera: capa de la atmósfera del sol que se encuentra sobre la fotosfera. (p. 109)

comet: space object made of dust and rock particles mixed with frozen water, methane, and ammonia that forms a bright coma as it approaches the Sun. (p. 90)

constellation: group of stars that forms a pattern in the sky that looks like a familiar object (Libra), animal (Pegasus), or character (Orion). (p. 104)

corona: outermost, largest layer of the Sun's atmosphere; extends millions of kilometers into space and has temperatures up to 2 million K. (p. 109)

cometa: objeto espacial formado por partículas de polvo y roca mezcladas con agua congelada, metano y amoníaco que forman una cola brillante cuando se aproxima al sol. (p. 90)

constelación: grupo de estrellas que forma un patrón en el cielo y que semeja un objeto (Libra), un animal (Pegaso) o un personaje familiar (Orión). (p. 104)

corona: capa más externa y más grande de la atmósfera solar; se extiende millones de kilómetros dentro del espacio y tiene una temperatura hasta de 2 millones de grados Kelvin. (p. 109)

E

Earth: third planet from the Sun; has an atmosphere that protects life and surface temperatures that allow water to exist as a solid, liquid, and gas. (p. 78)

electromagnetic spectrum: arrangement of electromagnetic waves according to their wavelengths. (p. 9)

ellipse (ee LIHPS): elongated, closed curve that describes Earth's yearlong orbit around the Sun. (p. 43)

equinox (EE kwuh nahks): twice-yearly time—each spring and fall—when the Sun is directly over the equator and the number of daylight and nighttime hours are equal worldwide. (p. 45)

Tierra: tercer planeta más cercano al sol; tiene una atmósfera que protege la vida y temperaturas en su superficie que permiten la presencia de agua en estado sólido, líquido y gaseoso. (p. 78)

espectro electromagnético: ordenamiento de las ondas electromagnéticas de acuerdo con su longitud de onda. (p. 9)

elipse: curva cerrada y elongada que describe la órbita anual de la Tierra alrededor del sol. (p. 43)

equinoccio: dos veces al año—en primavera y otoño—cuando el sol está posicionado directamente sobre el ecuador y el número de horas del día y de la noche son iguales en todo el mundo. (p. 45)

F

full moon: phase that occurs when all of the Moon's surface facing Earth reflects light. (p. 47)

luna llena: fase que ocurre cuando toda la superficie de la luna frente a la Tierra refleja la luz del sol. (p. 47)

G

galaxy: large group of stars, dust, and gas held together by gravity; can be elliptical, spiral, or irregular. (p. 120)

giant: late stage in the life of comparatively low-mass main sequence star in which hydrogen in the core is deleted, the core contracts and temperatures inside the star increase, causing its outer layers to expand and cool. (p. 117)

Great Red Spot: giant, high-pressure storm in Jupiter's atmosphere. (p. 82)

galaxia: grupo grande de estrellas, polvo y gas en donde todo está unido por gravedad; puede ser elíptica, espiral o irregular. (p. 120)

gigante: etapa tardía en la vida de una estrella de secuencia principal, de relativamente poca masa, en la que el hidrógeno en el núcleo está agotado, el núcleo se contrae y la temperatura en el interior de la estrella aumenta, causando que las capas externas se expandan y enfríen. (p. 117)

La Gran Mancha Roja: tormenta gigante de alta presión en la atmósfera de Júpiter. (p. 82)

I

impact basin: a hollow left on the surface of the Moon caused by an object striking its surface. (p. 57)

cráter de impacto: un hueco dejado en la superficie de la luna causada por un objeto que chocó contra su superficie. (p. 57)

J

Jupiter: largest and fifth planet from the Sun; contains more mass than all the other planets combined, has continuous storms of high-pressure gas, and an atmosphere mostly of hydrogen and helium. (p. 82)

light-year: unit representing the distance light travels in one year—about 9.5 trillion km—used to record distances between stars and galaxies. (p. 107)

lunar eclipse: occurs when Earth's shadow falls on the Moon. (p. 50)

Júpiter: el quinto planeta más cercano al sol, y también el más grande; contiene más masa que todos los otros planetas en conjunto, tiene tormentas continuas de gas a alta presión y una atmósfera compuesta principalmente por hidrógeno y helio. (p. 82)

año luz: unidad que representa la distancia que la luz viaja en un año—cerca de 9.5 trillones de kilómetros—usada para registrar las distancias entre las estrellas y las galaxias. (p. 107)

eclipse lunar: ocurre cuando la sombra de la Tierra cubre la luna. (p. 50)

M

maria (MAHR ee uh): dark-colored, relatively flat regions of the Moon formed when ancient lava reached the surface and filled craters on the Moon's surface. (p. 51)

Mars: fourth planet from the Sun; has polar ice caps, a thin atmosphere, and a reddish appearance caused by iron oxide in weathered rocks and soil. (p. 78)

Mercury: smallest planet, closest to the Sun; does not have a true atmosphere; has a surface with many craters and high cliffs. (p. 76)

meteor: a meteoroid that burns up in Earth's atmosphere. (p. 91)

meteorite: a meteoroid that strikes the surface of a moon or planet. (p. 92)

moon phase: change in appearance of the Moon as viewed from the Earth, due to the relative positions of the Moon, Earth, and Sun. (p. 47)

mares: regiones de la Luna relativamente planas y de color oscuro que se formaron cuando la lava alcanzó la superficie y llenó los cráteres en la seperficie lunar. (p. 51)

Marte: cuarto planeta más cercano al sol; tiene casquetes de hielo polar, una atmósfera delgada y una apariencia rojiza causada por el óxido de hierro presente en las rocas y suelo de su superficie. (p. 78)

Mercurio: el planeta más pequeño y más cercano al sol; no tiene una atmósfera verdadera; tiene una superficie con muchos cráteres y grandes acantilados. (p. 76)

meteoro: un meteoroide que se incinera en la atmósfera de la Tierra. (p. 91)

meteorito: un meteoroide que choca contra la superficie de la luna o de algún planeta. (p. 92)

fase lunar: cambio en la apariencia de la luna según es vista desde la Tierra; se debe a las posiciones relativas de la luna, la Tierra y el sol. (p. 47)

N

nebula: large cloud of gas and dust that contracts under gravitational force and breaks apart into smaller pieces, each of which might collapse to form a star. (p. 116)

nebulosa: nube grande de polvo y gas que se contrae bajo la fuerza gravitacional y se descompone en pedazos más pequeños, cada uno de los cuales se puede colapsar para formar una estrella. (p. 116)

Neptune: usually the eighth planet from the Sun; is large and gaseous, has rings that vary in thickness, and is bluish-green in color. (p. 86)

neutron star: collapsed core of a supernova that can shrink to about 20 km in diameter and contains only neutrons in the dense core. (p. 118)

new moon: moon phase that occurs when the Moon is between Earth and the Sun, at which point the Moon cannot be seen because its lighted half is facing the Sun and its dark side faces Earth. (p. 47)

Neptuno: el octavo planeta desde el sol; es grande y gaseoso, tiene anillos que varían en espesor y tiene un color verde-azulado. (p. 86)

estrella de neutrones: núcleo colapsado de una supernova que puede contraerse hasta tener un diámetro de 20 kilómetros y contiene sólo neutrones en su denso núcleo. (p. 118)

luna nueva: fase lunar que ocurre cuando la luna se encuentra entre la Tierra y el sol, punto en el cual la luna no puede verse porque su mitad iluminada está frente al sol y su lado oscuro frente a la Tierra. (p. 47)

observatory: building that can house an optical telescope; often has a dome-shaped roof that can be opened for viewing. (p. 10)

orbit: curved path followed by a satellite as it revolves around an object. (p. 17)

observatorio: edificación que puede albergar un telescopio óptico; a menudo tiene un techo en forma de domo que puede abrirse para la observación. (p. 10)

órbita: trayectoria curva seguida por un satélite conforme gira alrededor de un objeto. (p. 17)

P

photosphere: lowest layer of the Sun's atmosphere; gives off light and has temperatures of about 6,000K. (p. 109)

Pluto: considered to be the ninth planet from the Sun; has a solid icy-rock surface and a single moon, Charon. (p. 87)

Project Apollo: final stage in the U.S. program to reach the Moon, in which Neil Armstrong was the first human to step onto the Moon's surface. (p. 22)

Project Gemini: second stage in the U.S. program to reach the Moon, in which an astronaut team connected with another spacecraft in orbit. (p. 21)

Project Mercury: first step in the U.S. program to reach the Moon; orbited a piloted spacecraft around Earth and brought it back safely. (p. 21)

fotosfera: capa más interna de la atmósfera del sol; emite luz y tiene temperaturas de cerca de 6,000 grados Kelvin. (p. 109)

Plutón: considerado como el noveno planeta desde el sol; tiene una superficie sólida de roca congelada y una luna, Caronte. (p. 87)

Proyecto Apolo: etapa final en el proyecto norteamericano para llegar a la luna en el que Neil Armstrong fue el primer ser humano en caminar sobre la superficie lunar. (p. 22)

Proyecto Géminis: segunda etapa del proyecto norteamericano para llegar a la luna en el que un grupo de astronautas se conectó con otra nave espacial en órbita. (p. 21)

Proyecto Mercurio: primera etapa del proyecto norteamericano para llegar a la luna en el que una nave espacial tripulada recorrió la órbita de la Tierra y regresó de manera segura. (p. 21)

R

radio telescope: collects and records radio waves traveling through space; can be used day or night under most weather conditions. (p. 13)

radiotelescopio: recolecta y registra ondas de radio que viajan a través del espacio; puede usarse de día o de noche en la mayoría de condiciones climáticas. (p. 13)

reflecting telescope: optical telescope that uses a concave mirror to focus light and form an image at the focal point. (p. 10)

refracting telescope: optical telescope that uses a double convex lens to bend light and form an image at the focal point. (p. 10)

revolution: Earth's yearlong elliptical orbit around the Sun. (p. 43)

rocket: special engine that can work in space and burns liquid or solid fuel. (p. 15)

rotation: spinning of Earth on its imaginary axis, which takes about 24 hours to complete and causes day and night to occur. (p. 41)

telescopio reflectante: telescopio óptico que utiliza un espejo cóncavo para enfocar la luz y formar una imagen en el punto focal. (p. 10)

telescopio de refracción: telescopio óptico que utiliza un lente doble convexo para formar una imagen en el punto focal. (p. 10)

revolución: órbita elíptica de un año de duración que la Tierra recorre alrededor del sol. (p. 43)

cohete: máquina especial que puede funcionar en el espacio y quema combustible sólido o líquido. (p. 15)

rotación: rotación de la Tierra sobre su eje imaginario, lo cual toma cerca de 24 horas para completarse y causa la alternancia entre el día y la noche. (p. 41)

S

satellite: any natural or artificial object that revolves around another object. (p. 17)

Saturn: second-largest and sixth planet from the Sun; has a complex ring system, at least 31 moons, and a thick atmosphere made mostly of hydrogen and helium. (p. 84)

solar eclipse: occurs when the Moon passes directly between the Sun and Earth and casts a shadow over part of Earth. (p. 49)

solar system: system of nine planets, including Earth, and other objects that revolve around the Sun. (p. 71)

solstice: twice-yearly point at which the Sun reaches its greatest distance north or south of the equator. (p. 44)

space probe: instrument that travels far into the solar system and gathers data to send back to Earth. (p. 18)

space shuttle: reusable spacecraft that can carry cargo, astronauts, and satellites to and from space. (p. 23)

space station: large facility with living quarters, work and exercise areas, and equipment and support systems for humans to live and work in space and conduct research. (p. 24)

sphere (SFIHR): a round, three-dimensional object whose surface is the same distance from its center at all points; Earth is a sphere that bulges somewhat at the equator and is slightly flattened at the poles. (p. 40)

satélite: cualquier objeto natural o artificial que gire alrededor de otro objeto. (p. 17)

Saturno: además de ser el sexto planeta más cercano al sol, también es el segundo en tamaño; tiene un sistema de anillos complejo, por lo menos 31 lunas y una atmósfera gruesa compuesta principalmente de hidrógeno y helio. (p. 84)

eclipse solar: ocurre cuando la luna pasa directamente entre el sol y la Tierra y se genera una sombra sobre una parte de la Tierra. (p. 49)

sistema solar: sistema de nueve planetas, incluyendo a la Tierra y otros objetos que giran alrededor del sol. (p. 71)

solsticio: punto en el cual dos veces al año el sol alcanza su mayor distancia al norte o al sur del ecuador. (p. 44)

sonda espacial: instrumento que viaja grandes distancias en el sistema solar, recopila datos y los envía a la Tierra. (p. 18)

trasbordador espacial: nave espacial reutilizable que puede llevar carga, astronautas y satélites hacia y desde el espacio. (p. 23)

estación espacial: instalación grande con áreas para hospedarse, trabajar y hacer ejercicio; tiene equipos y sistemas de apoyo para que los seres humanos vivan, trabajen y lleven a cabo investigaciones en el espacio. (p. 24)

esfera: un objeto tridimensional y redondo donde cualquier punto de su superficie está a la misma distancia del centro; la Tierra es una esfera algo abultada en el ecuador y ligeramente achatada en los polos. (p. 40)

sunspots: areas on the Sun's surface that are cooler and less bright than surrounding areas, are caused by the Sun's magnetic field, and occur in cycles. (p. 110)

supergiant: late stage in the life cycle of a massive star in which the core heats up, heavy elements form by fusion, and the star expands; can eventually explode to form a supernova. (p. 119)

manchas solares: áreas en la superficie solar que son más frías y menos brillantes que las áreas circundantes, son causadas por el campo magnético solar y ocurren en ciclos. (p. 110)

supergigante: etapa tardía en el ciclo de vida de una estrella masiva en la que el núcleo se calienta, se forman elementos pesados por fusión y la estrella se expande; eventualmente puede explotar para formar una supernova. (p. 119)

Uranus (YOOR uh nus): seventh planet from the Sun; is large and gaseous, has a distinct bluish-green color, and rotates on an axis nearly parallel to the plane of its orbit. (p. 85)

Urano: séptimo planeta desde el sol; es grande y gaseoso, tiene un color verde-azulado distintivo y gira sobre un eje casi paralelo al plano de su órbita. (p. 85)

Venus: second planet from the Sun; similar to Earth in mass and size; has a thick atmosphere and a surface with craters, faultlike cracks, and volcanoes. (p. 77)

Venus: segundo planeta más cercano al sol; similar a la Tierra en masa y tamaño; tiene una atmósfera gruesa y una superficie con cráteres, grietas similares a fallas y volcanes. (p. 77)

waning: describes phases that occur after a full moon, as the visible lighted side of the Moon grows smaller. (p. 48)

waxing: describes phases following a new moon, as more of the Moon's lighted side becomes visible. (p. 48)

white dwarf: late stage in the life cycle of a comparatively low-mass main sequence star; formed when its core depletes its helium and its outer layers escape into space, leaving behind a hot, dense core. (p. 118)

menguante: describe las fases posteriores a la luna llena, de manera que el lado iluminado de la luna es cada vez menos visible. (p. 48)

creciente: describe las fases posteriores a la luna nueva, de manera que el lado iluminado de la luna es cada vez más visible. (p. 48)

enana blanca: etapa tardía en el ciclo de vida de una estrella de secuencia principal, de relativamente poca masa, formada cuando el núcleo agota su helio y sus capas externas escapan al espacio, dejando atrás un núcleo denso y caliente. (p. 118)

Italic numbers = illustration/photo   **Bold numbers = vocabulary term**
lab = a page on which the entry is used in a lab
act = a page on which the entry is used in an activity

Index

Magnification Key: Magnifications listed are the magnifica-
tions at which images were originally photographed.
LM–Light Microscope
SEM–Scanning Electron Microscope
TEM–Transmission Electron Microscope

Acknowledgments: Glencoe would like to acknowledge the
artists and agencies who participated in illustrating this pro-
gram: Absolute Science Illustration; Andrew Evansen; Argosy;
Articulate Graphics; Craig Attebery represented by Frank &
Jeff Lavaty; CHK America; John Edwards and Associates;
Gagliano Graphics; Pedro Julio Gonzalez represented by
Melissa Turk & The Artist Network; Robert Hynes repre-
sented by Mendola Ltd.; Morgan Cain & Associates; JTH
Illustration; Laurie O'Keefe; Matthew Pippin represented by
Beranbaum Artist's Representative; Precision Graphics;
Publisher's Art; Rolin Graphics, Inc.; Wendy Smith repre-
sented by Melissa Turk & The Artist Network; Kevin Torline
represented by Berendsen and Associates, Inc.; WILDlife
ART; Phil Wilson represented by Cliff Knecht Artist
Representative; Zoo Botanica.

Photo Credits

Cover (tl)NASA/Science Photo Library/Photo Researchers,
(tr)Billy & Sally Fletcher/Tom Stack & Assoc., (b)Photodisc;
i ii (tl)NASA/Science Photo Library/Photo Researchers,
(tr)Billy & Sally Fletcher/Tom Stack & Assoc., (b)Photodisc;
iv (bkgd)John Evans, cover: (tl)NASA/Science Photo Library/
Photo Researchers, (tr)Billy & Sally Fletcher/Tom Stack &
Assoc., (b)Photodisc; v (t)PhotoDisc, (b)John Evans
vi (l)John Evans, (r)Geoff Butler; vii (l)John Evans,
(r)PhotoDisc; viii PhotoDisc; ix Aaron Haupt Photography;
x (t)Julian Baum/Science Photo Library/Photo Researchers,
(b)NASA/Science Photo Library/Photo Researchers; xi AFP/
CORBIS; xii NASA; 1 Pekka Parviainen/Science Photo Library/
Photo Researchers; 2 (t)David J. Phillip/AP/Wide World
Photos; 2–3 Malin Space Science Systems/NASA; 3 (t)Malin
Space Science Systems/NASA/JPL, (br)courtesy DC Golden;
5 Geco UK/Science Photo Library/Photo Researchers;
6–7 TSADO/NASA/Tom Stack & Assoc.; 8 (l)Weinberg-Clark/
The Image Bank/Getty Images, (r)Stephen Marks/The Image
Bank/Getty Images; 9 (l)PhotoEdit, Inc., (r)Wernher Krutein/
Liaison Agency/Getty Images; 10 Chuck Place/Stock Boston;
11 NASA; 12 (t)Roger Ressmeyer/CORBIS, (b)Simon Fraser/
Science Photo Library/Photo Researchers; 13 Raphael
Gaillarde/Liaison Agency/Getty Images; 14 (t)Icon Images,
(b)Diane Graham-Henry & Kathleen Culbert-Aguilar;
15 NASA; 16 NASA/Science Photo Library/Photo Researchers;
17 NASA; 18 (Mariner 2, Pioneer 10)NASA/Science Source/
Photo Researchers, (Viking 1)M. Salaber/Liaison Agency/
Getty Images, (Magellan)Julian Baum/Science Photo Library/
Photo Researchers; 19 (Venera 8)Dorling Kindersley Images,
(Surface of Venus)TASS from Sovfoto, (Mercury, Venus)
NASA/JPL, (Voyager 2, Neptune)NASA/JPL/Caltech, (others)
NASA; 20 AFP/CORBIS; 21 NASA; 22 NASA/Science Source/
Photo Researchers; 23 NASA/Liaison Agency/Getty Images;
24 (t)NASA, (b)NASA/Liaison Agency/Getty Images;
25 NASA/Science Source/Photo Researchers; 26 NASA/JPL/
Malin Space Science Systems; 27 NASA/JPL/Liaison Agency/
Getty Images; 28 (t)David Ducros/Science Photo Library/
Photo Researchers, (b)NASA; 29 The Cover Story/CORBIS;
30 Roger Ressmeyer/CORBIS; 31 Doug Martin; 32 Robert
McCall; 33 (l)Novosti/Science Photo Library/Photo
Researchers, (c)Roger K. Burnard, (r)NASA; 36 Tom Steyer/
Getty Images; 37 NASA/Science Photo Library/Photo
Researchers; 38–39 Chad Ehlers/Stone/Getty Images;
48 (bl)Richard J. Wainscoat/Peter Arnold, Inc., (others)Lick
Observatory; 50 Dr. Fred Espenak/Science Photo Library/
Photo Researchers; 51 Bettmann/CORBIS; 52 NASA;
54 Roger Ressmeyer/CORBIS; 57 BMDO/NRL/LLNL/Science
Photo Library/Photo Researchers; 58 (t)Zuber et al/Johns
Hopkins University/NASA/Photo Researchers, (b)NASA;
59 NASA; 61 Matt Meadows; 62 Cosmo Condina/Stone;
64 Lick Observatory; 65 NASA; 68–69 Roger Ressmeyer/
CORBIS; 69 Matt Meadows; 72 European Southern
Observatory/Photo Researchers; 74 Bettmann/CORBIS;
76 USGS/Science Photo Library/Photo Researchers;
77 (t)NASA/Photo Researchers, (b)JPL/TSADO/Tom Stack &
Assoc.; 78 (t)Science Photo Library/Photo Researchers,
(bl)USGS/TSADO/Tom Stack & Assoc., (bc)USGS/Tom Stack
& Assoc., (br)USGS/Tom Stack & Assoc.; 79 NASA/JPL/
Malin Space Science Systems; 81 Science Photo Library/
Photo Researchers; 82 (l)NASA/Science Photo Library/Photo
Researchers, (r)CORBIS; 83 (Io)USGS/TSADO/Tom Stack &
Assoc., (Europa)NASA/JPL/Photo Researchers, (Ganymede)
NASA/TSADO/Tom Stack & Assoc., (Callisto)JPL, (b)NASA;
84 JPL; 85 Heidi Hammel/NASA; 86 (l)NASA/Science
Source/Photo Researchers, (r)NASA/JPL/TSADO/Tom Stack
& Assoc.; 87 CORBIS; 88 (Mercury)NASA/JPL/TSADO/Tom
Stack & Assoc., (Venus)NASA/Science Source/Photo
Researchers, (Earth)CORBIS, (Mars)NASA/USGS/TSADO/
Tom Stack & Assoc.; 89 (Jupiter)NASA/Science Photo
Library/Photo Researchers, (Saturn)NASA/Science Source/
Photo Researchers, (Uranus)ASP/Science Source/Photo
Researchers, (Neptune)W. Kaufmann/JPL/Science Source/
Photo Researchers, (Pluto)CORBIS; 90 Pekka Parviainen/
Science Photo Library/Photo Researchers; 91 Pekka
Parviainen/Science Photo Library/Photo Researchers;
92 Georg Gerster/Photo Researchers; 93 JPL/TSADO/Tom
Stack & Assoc.; 95 Bettmann/CORBIS; 96 (t b)Museum of
Natural History/Smithsonian Institution; 97 (t)NASA,
(bl)JPL/NASA, (br)file photo; 98 NASA/Science Source/
Photo Researchers; 100 John R. Foster/Photo Researchers;
102–103 TSADO/ESO/Tom Stack & Assoc.; 107 Bob
Daemmrich; 110 (t)Carnegie Institution of Washington,
(b)NSO/SEL/Roger Ressmeyer/CORBIS;111 (l)NASA,
(r)Picture Press/CORBIS, (b)Bryan & Cherry Alexander/
Photo Researchers; 112 Celestial Image Co./Science Photo
Library/Photo Researchers; 113 Tim Courlas; 115 Luke
Dodd/Science Photo Library/Photo Researchers;
118 AFP/CORBIS; 119 NASA; 121 (t)Kitt Peak National
Observatory, (b)CORBIS; 125 R. Williams (ST ScI)/NASA;
126 Matt Meadows; 128 Dennis Di Cicco/Peter Arnold, Inc.;
129 (l)file photo, (r)AFP/CORBIS; 134 PhotoDisc; 136 Tom
Pantages; 140 Michell D. Bridwell/PhotoEdit, Inc.;
141 (t)Mark Burnett, (b)Dominic Oldershaw; 142 StudiOhio;
143 Timothy Fuller; 144 Aaron Haupt; 146 KS Studios;
147 Matt Meadows; 150 Amanita Pictures; 151 Bob
Daemmrich; 153 Davis Barber/PhotoEdit, Inc.; 169 Matt
Meadows; 170 (l)Dr. Richard Kessel, (c)NIBSC/Science
Library/Photo Researchers, (r)David John/Visuals Unlimited;
171 (t)Runk/Schoenberger from Grant Heilman, (bl)Andrew
Syred/Science Photo Library/Photo Researchers, (br)Rich
Brommer; 172 (tr)G.R. Roberts, (l)Ralph Reinhold/Earth
Scenes, (br)Scott Johnson/Animals Animals; 173 Martin
Harvey/DRK Photo.

PERIODIC TABLE OF THE ELEMENTS

Columns of elements are called groups. Elements in the same group have similar chemical properties.

Gas
Liquid
Solid
Synthetic

Element — Hydrogen
Atomic number — 1
Symbol — H
Atomic mass — 1.008
State of matter

The first three symbols tell you the state of matter of the element at room temperature. The fourth symbol identifies elements that are not present in significant amounts on Earth. Useful amounts are made synthetically.

Group	1	2	3	4	5	6	7	8	9
1	Hydrogen 1 H 1.008								
2	Lithium 3 Li 6.941	Beryllium 4 Be 9.012							
3	Sodium 11 Na 22.990	Magnesium 12 Mg 24.305							
4	Potassium 19 K 39.098	Calcium 20 Ca 40.078	Scandium 21 Sc 44.956	Titanium 22 Ti 47.867	Vanadium 23 V 50.942	Chromium 24 Cr 51.996	Manganese 25 Mn 54.938	Iron 26 Fe 55.845	Cobalt 27 Co 58.933
5	Rubidium 37 Rb 85.468	Strontium 38 Sr 87.62	Yttrium 39 Y 88.906	Zirconium 40 Zr 91.224	Niobium 41 Nb 92.906	Molybdenum 42 Mo 95.94	Technetium 43 Tc (98)	Ruthenium 44 Ru 101.07	Rhodium 45 Rh 102.906
6	Cesium 55 Cs 132.905	Barium 56 Ba 137.327	Lanthanum 57 La 138.906	Hafnium 72 Hf 178.49	Tantalum 73 Ta 180.948	Tungsten 74 W 183.84	Rhenium 75 Re 186.207	Osmium 76 Os 190.23	Iridium 77 Ir 192.217
7	Francium 87 Fr (223)	Radium 88 Ra (226)	Actinium 89 Ac (227)	Rutherfordium 104 Rf (261)	Dubnium 105 Db (262)	Seaborgium 106 Sg (266)	Bohrium 107 Bh (264)	Hassium 108 Hs (277)	Meitnerium 109 Mt (268)

The number in parentheses is the mass number of the longest-lived isotope for that element.

Rows of elements are called periods. Atomic number increases across a period.

The arrow shows where these elements would fit into the periodic table. They are moved to the bottom of the table to save space.

Lanthanide series	Cerium 58 Ce 140.116	Praseodymium 59 Pr 140.908	Neodymium 60 Nd 144.24	Promethium 61 Pm (145)	Samarium 62 Sm 150.36
Actinide series	Thorium 90 Th 232.038	Protactinium 91 Pa 231.036	Uranium 92 U 238.029	Neptunium 93 Np (237)	Plutonium 94 Pu (244)